Darwinism is the Doorway to Atheism

Why Creationists Become Evolutionists

55 Case Studies

by Jerry Bergman, Ph.D.

and

Kevin H. Wirth, Senior Editor and Contributor

Leafcutter Press
Your Insight to the World

Leafcutter Press, Publisher
Spokane Valley, WA 99037.
Published in the United States of America.

No portion of this book may be reproduced or transmitted in any form or by any means electronic or mechanical including photocopying, reprinting, or any information storage or retrieval system without a legitimately obtained license or permission in writing from Leafcutter Press, LLC.

Publisher's Note:
This book is Volume II of the "Doorway" series. The other volume in this series is titled:

Science is the Doorway to the Creator (Volume I) May 2019

Copyright Notice

Copyright © 2019 by Leafcutter Press. All Rights Reserved.

First Edition – November 2019

Library Catalog Data:

Bergman, Jerry R. (1946 –) Author
Wirth, Kevin (1951 -) Sr. Editor and Contributor
ISBN 978-0-9997992-1-5 (paperback)
1. Religious Discrimination
2. Discrimination – United States
3. Evolution
4. Academic freedom – United States
5. Creationism and Intelligent Design

Typography by Guy Forsythe of Crying Rocks LLC (Cryingrocks.com)

Books by Dr. Bergman on the Creation-Evolution issue

1. *Persuaded by the Evidence (editor).* Green Forest, AR: Master Books. 2008. 288 pages. With Doug Sharp.

2. *Slaughter of the Dissidents: The Shocking Truth About Killing the Careers of Darwin Doubters.* (Volume II of the *Slaughter of the Dissidents* trilogy). Southworth, WA: Leafcutter Press. 2008. 2nd revised edition published in 2012. With Kevin Wirth, editor and contributor.

3. *Silencing the Darwin Skeptics: The War Against Theists* (Volume II of the *Slaughter of the Dissidents* trilogy). Southworth, WA: Leafcutter Press. 2016. With Kevin Wirth, editor and contributor.

4. *Censoring the Darwin Skeptics. How Belief in Evolution is Enforced by Eliminating Dissidents.* (Volume III of the Slaughter of the Dissidents trilogy). Southworth, WA: Leafcutter Press. 2018. With Kevin Wirth, editor and contributor.

5. *The Dark Side of Darwin.* Green Forest, AR: New Leaf Press. 2011. Revised edition published in 2015.

6. *Hitler and the Nazi Darwinian Worldview: How the Nazi Eugenic Crusade for a Superior Race Caused the Greatest Holocaust in World History.* 2012. Kitchener, Ontario, Canada: Joshua Press.

7. *Transformed by the Evidence* edited with Doug Sharp. 2014. Southworth, WA: Leafcutter Press

8. *The Darwin Effect. Its influence on Nazism, Eugenics, Racism, Communism, Capitalism & Sexism.* 2014. Green Forest, AR: Master Books.

9. *Debunking Human Evolution Taught in Our Public Schools: A Guidebook for Christian Students, Parents, and Pastors.* Dec 27, 2015 with Dr. Daniel A. Biddle and David A. Bisbee. Genesis Apologetics Press. Folsom, CA.

10. *Debunking Human Evolution Taught in Our Public Schools: A Guidebook for Christian Students, Parents, and Pastors.* with Dr. Daniel A. Biddle and David A. Bisbee. Genesis Apologetics Press. Folsom, CA. 2nd edition. 2016.

11. *C. S. Lewis: Anti-Darwinist: A Careful Examination of the Development of His Views on Darwinism.* Eugene Oregon: Wipf & Stock Publishers. 2016.

12. *How Darwinism Corrodes Morality: Darwinism, Immorality, Abortion and the Sexual Revolution.* 2017. Kitchener, Ontario, Canada: Joshua Press.

13. *Evolution's Blunders, Frauds and Forgeries.* Atlanta, GA: CMI Publishing. 2017.

14. *Evolution is the Doorway to Atheism.* Spokane, WA: Leafcutter Press. 2019.

15. *Science is the Doorway to the Creator: Nobel Laureates and other Eminent Scientists who reject Orthodox Darwinism.* 2019. Southworth, WA: Leafcutter Press.

16. *Fossil Forensics: Separating Fact from Fantasy in Paleontology.* 2017. Tulsa, Oklahoma: Bartlett Publishing. 356 pages.

17. *The Three Major Pillars of Darwinism Demolished.* Tulsa, Oklahoma: Bartlett Publishing. 2019.

18. *Useless Organs: The Rise and Fall of the a Central Claim of Evolution.* Tulsa, Oklahoma: Bartlett Publishing. 2019.

19. *Poor Design An Invalid Argument Against Intelligent Design.* Tulsa, Oklahoma: Bartlett Publishing. 2019.

20. *God in Eisenhower's Life, Military Career, and Presidency.* Eugene, Oregon: Wipf & Stock Publishers. 2019.

21. *How Germany and the United States Produced the Holocaust.* Kitchener, Ontario: Joshua Press Expect in 2019.

22. *History of the United Methodist Church's Opposition to Creationism and Intelligent Design.* 2019. Southworth, WA: Leafcutter Press (in press).

23. *The Minor Pillars of Darwinism Demolished.* Tulsa, Oklahoma: Bartlett Publishing. 2019.

Acknowledgements

The many persons who read and commented on this manuscript in whole or part include Dr. Raymond Damadian, M.D., Michael Dennis, Ph.D., Peter Lassen, M.Sc.E., Shelley Hausch, Ellen Yeske, Dr. David Herbert, Steven E. Woodworth Ph.D. Associate Professor of History Texas Christian University, Professor Tom McMullen, Department of History, Georgia Southern University, Statesboro, GA, Wayne Frair, Ph.D. Professor Emeritus at The Kings College, New York, Jan Peczkis, M.A. and Kevin H. Wirth, Publisher at Leafcutter Press. I also want to thank late Albert Szent-Györgyi, Ph.D., Theodore J. Siek Ph.D., Bert Thompson, Ph.D., John UpChurch, Mary Ann Stuart, M.A., Bryce Gaudian and Clifford Lillo, M.A. and the many reviewers.

Table of Contents

Introduction by Bruce Malone xvii
Publisher's Introduction by Kevin H. Wirth xix
Author's Introduction by Dr. Jerry Bergman xxix
Chapter 1 Charles Darwin 1
Chapter 2 Professor Warder Clyde Allee, Ph.D. 5
Chapter 3 Brian Alters, Ph.D. 9
Chapter 4 Hector Avalos, Ph.D. 13
Chapter 5 Rev. Francisco J. Ayala, Ph.D. 19
Chapter 6 Edward G. Babinski 25
Chapter 7 Robert Bakker, Ph.D. 27
Chapter 8 Dan Barker 31
Chapter 9 Henri Bergson 35
Chapter 10 Ross Blythe 45
Chapter 11 Bart Campolo 51
Chapter 12 Richard Colling, Ph.D. 55
Chapter 13 Francis Crick, Ph.D. 61
Chapter 14 Raymond Dart, Ph.D. 65
Chapter 15 Richard Dawkins, Ph.D. 67
Chapter 16 Eugene Dubois, Ph.D. 73
Chapter 17 David Duke 77
Chapter 18 Albert Einstein 93
Chapter 19 Rachel Held Evans 97
Chapter 20 Martin Gardner 101
Chapter 21 Karl W. Giberson, Ph.D. 105
Chapter 22 Philip Dean Gingerich, Ph.D. 111
Chapter 23 Stephen Godfrey, Ph.D. 113
Chapter 24 Hiram Bentley Glass, Ph.D. 121
Chapter 25 Richard Goldschmidt, Ph.D. 123
Chapter 26 Ernst Haeckel, Ph.D. 127
Chapter 27 Stephen Hawking, Ph.D. 129
Chapter 28 Kevin R. Henke, Ph.D. 147

Chapter 29 Denis Lamoureau, D.D.S., Ph.D.. 151
Chapter 30 Louis Leakey, Ph.D.. 157
Chapter 31 Glenn Robert Morton 159
Chapter 32 David Mills. 163
Chapter 33 PZ Myers, Ph.D. 165
Chapter 34 Ronald Numbers, Ph.D. 171
Chapter 35 George Perdikis. 175
Chapter 36 William Provine, Ph.D.. 177
Chapter 37 Chet Rayno, Ph.D. 181
Chapter 38 Stanley Arthur Rice, Ph.D. 193
Chapter 39 Alan Rogers, Ph.D 197
Chapter 40 Nicolas Adriano's Rupke, Ph.D. 199
Chapter 41 Michael Shermer, Ph.D. 203
Chapter 42 George Gaylord Simpson, Ph.D. 207
Chapter 43 Howard N. Teeple 209
Chapter 44 Charles Templeton. 211
Chapter 45 Howard Van Till, Ph.D.. 213
Chapter 46 Dennis Venema, Ph.D.. 217
Chapter 47 Geerat Vermeij, Ph.D. 235
Chapter 48 Jason Wiles, Ph.D. 239
Chapter 49 H. G. Wells 243
Chapter 50 Edward O. Wilson, Ph.D. 247
Chapter 51 Davis Young, Ph.D.. 251
Chapter 52 Frank R. Zindle 255
Chapter 53 Rev. John Zingaro 259
Chapter 54 Joshua Zorn, Ph.D.. 263
Chapter 55 Libby Anne 271
Chapter 56 Conclusions 281
Appendix A – Turning Point Topics 287
Appendix B – Video and Other Media Resource Providers 309
Appendix C – Apologists for Evidence of a Designed Universe 311

Endorsement

If Darwin never was born, Bergman documents in his writings, the world would very likely be a very different place today. No WWI or WWII, no Communism, no Korean or Vietnam War, no Holocaust, no eugenics movement and no Cold War. Karl Marx would have continued to write excellent Christian literature and may have become another C S Lewis. Joseph Stalin would have become a Priest, and maybe even a bishop, and a leader in the Russian Orthodox Church. Einstein would have been a Messianic Jew still preaching the Christian gospel of peace and doing his science as a Professor emeritus at the University of Berlin. Hitler would have finally received recognition for his artwork and lived his dream of becoming a great artist, although his pictures still tended to be void of people. Darwin himself would have become a country parson, and, his health very good, very active in the church and widely recognized as a devoted family man. Germany would still be the industrial leader of Europe, focusing on restoring its centuries old buildings in all of its major cities, including Berlin, Dresden, Cologne, Hamburg and dozens of others.

The 410,000 German civilians killed by Allied air raids would be grandparents today as would the over 20 million German soldiers killed in the war. The 80 million murdered by Stalin's rule and the 60 million by Mao's rule plus the almost 30 million, give or take a few million, by the other communists rulers would also be grandparents, many great grandparents. New surveys of the world's leading scientists would show 98 percent would be theists, mostly Christian, instead of 98 percent atheists as is the case today. Most of Europe would be active Christians in contrast to the miniscule number today. Bergman would still be teaching at Bowling Green State University and would be celebrating his 50th wedding anniversary next year. He would also have 4 doctorates instead of only two. Yes, the world would likely be a very different place. The claim that someone else would have created evolution is not likely because only Darwin had the money, influence, position and determination to murder (the word he used) God, and it is improbable with our much greater knowledge of science today that evolutionism would have any credibility.

Willard Lake, retired attorney

Dedication

To the ministers who claim "we need to preach Christ only. The creation concern is a side issue that is not important." This book will document that it is *very* important. It is the doorway to Atheism and one of the most critical issues in our culture today. And to Henry Morris III who during my 2018 employment at ICR allowed me to write, or in most cases complete, 19 books, all of which have now been accepted for publication after the normal extensive review process.

Select Quotes

I've come to see that the single most powerful force for dissolving religious faith in the West was, and still is, Darwinism.[1]

John Zmirak

"My attempts to demonstrate evolution by an experiment carried on for more than 40 years have completely failed. It is not even possible to make a caricature of an evolution out of paleobiological facts. The idea of an evolution rests on pure belief."[2]

Professor Nils Heribert Nilsson, noted Lund University evolutionist, botanist and geneticist.

---

1   Zmirak, John. 2018. 'Human Zoos' Exhibits the Racist Toll of Darwinism, *The Stream*, November 19. https://stream.org/human-zoos-exhibits-the-racist-toll-of-darwinism/
2   Nilsson, H. 1953. *Synthetische Artbildting* [sic. *Artbildung*]. Verlag CWK Gleerup, Lund, Sweden, p. 1185.

Note About the Assignment of Worldview

I have attempted to identify the worldview of people in the title for most chapters of this book. In some cases, it was not clear if the person was an atheist, agnostic, simply a "none" or another label. It was in most cases clear though, that they were all creation oriented at one time or another before becoming evolutionists. While not everyone I report here became an atheist, they all shifted from a creation worldview to an evolutionary one, which is on the slippery continuum towards atheism. I fully recognize that many Theistic Evolutionists will not end up as atheists, but the move away from a creation worldview is undeniably a shift towards atheism, since naturalistic evolution is a materialist concept without need of a creator. Some reported here continued to retain some semblance of creation views, but other individuals radically changed their worldview perspective. If any of my assessments about the people reported here are in error, I will be happy to make appropriate changes in the next edition of this book. I have consistently striven to be accurate above all, but may not have achieved this goal in every case.

The typical progression along the worldview continuum may include Young Earth Creationist (YEC) to Intelligent Design (ID), to Theistic Evolutionist (TE), to evolution with a veneer of theism, to a none, then agnostic, and lastly atheist, but it is not always obvious where someone is on this scale. Furthermore, many stop at various points on this continuum along the way. For example, many move from YEC to ID and remain there for the rest of their life.

This work is not intended to be a comprehensive examination of the reasons why most of the individuals covered in this book chose to leave their Christian faith behind, although Kevin Wirth delved into this somewhat in his introduction. Many of the chapters are simply intended to be a series of vignettes illustrating that the person discussed became convinced evolutionary explanations were more reasonable than a creationist perspective. For some, I go into much greater detail so that readers might get a feel for some of the behind-the-scenes activity in their lives in order to provide a broader view of events and people who influenced them, and their responses.

The whole point of this volume is to illustrate that many Christians are beguiled by the seemingly rational explanations offered by a Darwinian worldview, and that effective arguments to support a creation worldview are compelling, and only need broader exposure. The key focus for readers of this book is to help them contemplate the significance of this widespread apostasy, why it occurs, and how to effectively address it.

Dr. Jerry Bergman

Charles Darwin's 1859 publication of On the Origin of Species *was the greatest scripture-killer ever penned. The book demolished ... an entire series of biblical claims by demonstrating that purely naturalistic processes—evolution and natural selection—could explain patterns in nature previously explainable only by invoking a Great Designer.*[3]

Introduction by Bruce Malone

Over the last 20 years I have given thousands of lectures to over 100,000 people on the scientific evidence supporting a Biblical creation worldview. The most common concern is how to bring loved ones and friends back to the Christian truths which they were raised to believe. In many ways, this is analogous to asking how to undo a car wreck that has already happened.

The American public education system is 100 years down the road of leaving God out of any possible consideration when explaining the origin of the universe, the coded information within DNA, the extraordinary complexity of life, the vast gulf between different body structures, and the geology of our planet. Since 1948 the Supreme Court has institutionalized atheism (the separation of church and state) within our public education system and three generations of young Americans have had God totally removed from their educational experience.

Consequently, He is assumed to be either irrelevant, or non-existent. Interpretations of geological, biological, and cosmic evidence seeming to support evolutionary explanations are like a fire-hose torrent of information blasting away at student's faith while the church shoots a squirt gun of simplistic (and often inaccurate) justifications for why to believe the Bible. More often, church leaders join in rejecting the emphatically clear teaching of the first 11 chapters of the Bible...only to be surprised when the next generation does not take the rest of the Bible seriously.

Darwinism is the Doorway to Atheism is a lighthouse warning of the impending crash of young people's faith upon the rocky shore of our educational system. Few would not give their very lives to save their children from a car speeding toward them as they played on a highway. Yet over 80% of Christian children are rejecting any trust in the Bible by the time they leave college. In case study after case study, Dr. Bergman has documented how

---

3   Jerry Coyne. *Fact vs. Faith*. New York: Viking, 2015, p. 2.

leading proponents of Biblical truth turned their backs on God's Word once they learned the superficial evidence that seemingly supports evolution and ancient origins of humans.

Far more important than protecting your child from an impending physical danger is making sure they have an accurate and realistic defense against attacks on their eternal destiny. Darwinism is riddled with inaccuracies, misinterpretations of evidence, and superficial explanations which make sense on the surface, but often fail on closer examination. Dr. Bergman has done a masterful job of documenting 53 of the thousands of possible examples of how those raised to believe in Christianity and the Bible became blinded to its truths. In essentially every case this could have been prevented had these lost souls been given an adequate understanding of the problems with evolution before becoming exposed to its lure. Don't make the same mistake with your children!

Bruce Malone, Retired Research Leader for the Dow Chemical Corp., is now the Executive Director for *Search for the Truth Ministries*

Publisher's Introduction by Kevin H. Wirth

As I was preparing this book for publication, I began to see some patterns emerging that readers would do well to consider. I was struck (and stunned) by the number of many intelligent and well-known individuals presented by Dr. Bergman who early in life held creation oriented perspectives, but who ended up traveling down a road that radically changed their worldview as they encountered what they perceived to be unresolvable conflicts between evolution and their creationist beliefs. Dr. Bergman also reports on the impact this decision has had on many of those he presents, and the lives of others they in turn have influenced. That said, I would like to add a few comments about why I think many people included in this volume (as well as others) may have been seriously misled causing them to shift to an evolutionary worldview.

The 500 pound gorilla in the room is the question *Why?* Why do so many Creationists, especially younger ones, eventually change their worldview to accept evolution and often end up becoming atheists? What was it that caused the people reported here to do a re-evaluation and often a complete rejection of their creation worldview? Why were they persuaded to accept evolution even though it is widely conceded by many evolutionists that much of the tenuous evidence proposed for their theory is based on speculation, extrapolation, and conjecture? And why couldn't they understand the clear and obvious evidence (which even many prominent evolutionists admit) of design in the universe? Finally, why did other individuals, after struggling with this dilemma, come through it with their design oriented worldview intact and remain convinced that Neo-Darwinism does not stand up to close scrutiny?[4] The stories about this struggle are documented in many of the cases Dr. Bergman presents here and provide us with many crucial insights and answers to these "why" questions.

One of the major reasons why so many people adopt an evolutionary perspective is because the public reputation of science is perceived to be almost sterling. Those who practice in areas of evolutionary science are seen to be generally trustworthy. But is that view justified? I would suggest that often it is not, particularly among those who give precedence to enforcing an ideological agenda supporting naturalistic evolution regardless of any disconfirming

---

[4] For example, Walter ReMine, a former evolutionist, became convinced that evolution was not reliable. His experience is similar to my own. https://www.youtube.com/watch?v=mj2f9Py4vEQ

evidence. Not that science isn't an excellent way to study the world around us – I have no problem with that at all. And I certainly have no issue with the honest intent of many who work on evolution related projects – but I also think many of them are victims of pre-programmed perception issues such as confirmation bias and seeing what they expect to see in their research. All too often objectivity gets lost in evolutionary assumptions that have become built into practically every research program where evolution-related issues are perceived as likely or relevant.

For example, a team of workers in London who recently investigated the functions of the bacterial flagellum assumed that an evolutionary explanation would be forthcoming, even before they started working on their project.[5] And guess what? They saw what they expected to see: evolutionary implications everywhere. Though perhaps well intentioned, such an approach that begins with the assumption of an evolutionary outcome is flawed from the start due to built-in bias. I can think of several alternative explanations for the wonderful work this team did, and none of it assumes any evolutionary conclusions.

What I take issue with is the way many evolutionists make similar assumptions, resulting in deliberate or unconsciously manipulated scientific research that nuance the evidence so that it formulates a pre-determined evolutionary conclusion. The findings produced by such efforts are then promoted as reliable to scientists, educators and the man or woman on the street. The trust given to those who promote the results of evolution-related projects and studies must be free from presuppositional bias, but unfortunately this is often not the case. Conclusions that automatically assume evolution as the underlying outcome of a research project need to be emphasized as conjecture wherever appropriate if the treasured reputation of science is to be preserved.

Readers would do well to keep in mind that the conclusions presented by naturalistic and theistic evolutionists will almost always be skewed in favor of their worldview, *even if the hard evidence does not support it*. Many evolutionists either stretch or nuance the evidence to fit their worldview, overlook the influence of their own confirmation bias, or continue in their quest for evolutionary evidence despite innumerable inconvenient, recurring, and vexing "anomalies."[6] They confidently assume that what we don't fully un-

---

5   News Staff. 2018. "Biologists Trace Evolution of Bacterial Flagellar Motors." *Sci-News*. January 9. This post includes some very awesome illustrations and photos (cryo-electron microscopy) of flagellar motors. (note: the article mistakenly identifiesthis process as "cryo-election microscopy...) http://www.sci-news.com/biology/bacterial-flagellar-motors-05612.html
6   Gould, Stephen Jay. 2002. *The Structure of Evolutionary Theory*. Cambridge, MA: The Belknap Press

derstand about evolution today will eventually form a more complete picture in the future as the missing pieces fall together. This "we will get further confirmation later" approach is a key reason why so many evolution-based researchers enthusiastically continue on in their valiant quest to uncover more evidence to support their worldview. The enticement that their work will contribute towards a growing understanding of what they believe to be the truth about the natural world spurs them on. Here we come to perhaps one of the most damning violations of the trust that so many of us place in scientists who are supposed to be objective investigators.

The Lewontin Imperative

Many of the elite evolution-oriented consensus holders and key promoters of evolution support what I call the *Lewontin Imperative*, which is well defined by Harvard evolutionary biologist Richard Lewontin.

> …Our willingness to accept scientific claims that are against common sense is the key to an understanding of the real struggle between science and the supernatural. We take the side of science *in spite* of the patent absurdity of some of its constructs, *in spite* of its failure to fulfill many of its extravagant promises of health and life, *in spite* of the tolerance of the scientific community for unsubstantiated just-so stories, because we have a prior commitment, a commitment to materialism. It is not that the methods and institutions of science somehow compel us to accept a material explanation of the phenomenal world, but, on the contrary, that we are forced by our *a priori* adherence to material causes to create an apparatus of investigation and a set of concepts that produce material explanations, no matter how counter-intuitive, no matter how mystifying to the uninitiated. Moreover, that materialism is absolute, for we cannot allow a Divine Foot in the door. [7]

This is a fairly well-known, even iconic, quote creationists frequently refer to, but it's well worth taking a deeper look at exactly what Lewontin is revealing about bias in science rather than allowing the evidence lead where it may. The point is, what Lewontin reveals here is a deeply ingrained attitude

---

of Harvard University Press, p. 755. Gould refers to the reluctant admissions in 1959 by George Gaylord Simpson that "…one of the most important theoretical problems in the whole history of life: is the sudden appearance [of life]…" as a "category of troubling 'anomalies.'"
7 Lewontin, Richard. 1997. "Billions and Billions of Demons," *The New York Review*, January 9, p. 31.

and modus operandi that appears to be widely practiced by many Darwinists and others who agree with this credo.

The Lewontin Imperative reveals how to craft a rigged directional needle in an ideological compass, designed to ensure that all evidence always points to evolution, *no matter what*. He also discloses the motive to justify why many scientists think this approach is warranted, i.e., no evidence, no matter how compelling, can allow a 'Divine Foot" to enter the domain of science. The imperative makes no allowance for scientists to merely *point to the possibility* of a Designer, even when it's as obvious as water in a river. But does the end justify the means? Has this approach to presenting scientific evidence become a quality to be extolled within the scientific community? Does the Lewontin tactic encourage others to create an illusory assemblage of what is then promoted as scientific evidence? The answer is apparently "yes," because the practices proposed by the Imperative are ubiquitous today and constantly appear in peer-reviewed scientific journals and other highly respected publications. The Imperative also has the effect of inciting stringent vigilante enforcement actions taken by those who believe they are authorized to seek and destroy the careers of anyone who stands in opposition to the Darwinian worldview.[8] This is an unfortunate, but fully predictable, outcome of adherence to the Lewontin Imperative.

This approach does more damage to the scientific enterprise than any of the shrill denunciations and protests handed out by many Darwinians who rail on about the dangers of allowing any Darwin skeptic to be heard. Part of the reason why Lewontin's Imperative is so dangerous is because it cultivates a type of deliberately one-sided groupthink among many evolutionary biologists and paleontologists. While many evolutionary thought leaders proclaim the dangers of giving any platform to Darwin dissenters, they are busy implementing a strategy designed to ensure their views are typically seen as reliably confirming evolution, even when the evidence is "absurd," or made of "…just-so stories."[9] Ensuring that the needle on the scientific compass always points to evolution is their primary concern, "no matter how counter-intuitive…" or "…mystifying." In other words, a materialist worldview must always be supported above all else, *even if the evidence indicates otherwise*. This also reveals how science, at least in the domain of evolution, has been abused by advocates of this Imperative as both a vehicle and cover story designed to support mate-

---

8   Bergman, Jerry and Kevin Wirth (editor and contributor). 2008-2018. *Slaughter of the Dissidents* (trilogy). Southworth, WA: Leafcutter Press.
9   I'm quoting directly from Lewontin's Imperative here.

rialism above all, rather than promote objective scientific discovery.

In almost any other enterprise, this approach would be clearly recognized as rigging the system, and dishonestly tipping the scales in their favor. Promoting evolution, or any theory for that matter, in the manner proscribed by Lewontin, seeks to manipulate others to into accepting a massaged understanding of what is then promoted as scientific reality. The Imperative encourages evidence tampering on a grand scale and is nothing less than incitement to commit fraud. Yet this approach has become a common, almost unconscious and routine exercise among many researchers, scientists, pundits, media reporters, and academics who promote what they call scientific evidence favoring evolution. And, many of them are dedicated to enforcing the Lewontonian maxims to the hilt, often using scandalous discrimination tactics.

My research indicates that the Lewontin Imperative ensures far more than just influencing an ideological shift. What it says is, even if science leads us to conclude that the universe is designed, scientists must never admit it. But where the Lewontin Imperative goes overboard is the collateral damage caused by this approach: the targeting and persecution of Darwin skeptics who dare to voice their opposition to the results of employing the Lewontin approach. Of particular interest to many Lewontin acolytes are Darwin skeptics who are perceived to be 'religious' in any way and are critical of evolutionary propositions, even if those notions are admittedly conjectural by leading evolutionists. Darwin dissidents are frequently targeted for removal from any role or influence they may have in either practicing or teaching about science.[10] This is because they pose a serious threat to exposing the tainted nature of scientific investigation promoted by innumerable scientists who agree with and adopt the Imperative.

We should consider how the Lewontin Imperative has been applied in various disciplines of science as I lay out this charge of hijacking the reputation of science to support what the Lewonton Imperative demands: a promote-at-all-costs materialist worldview. I will leave the bulk of identifying how Lewontin's prescription is implemented throughout the scientific enterprise for other workers to document, but will offer one extraordinary example for readers to consider. In my opinion we need look no further than how the evidence of fossils has been treated. Paleontologists admit to two very glaring and disconfirming pieces of evidence they've had to wrestle with.

First is the abrupt appearance of thousands of species that show up

---

10 Bergman, Jerry and Kevin Wirth (editor and contributor). 2018. *Censoring the Darwin Skeptics* (Vol III in the *Slaughter of the Dissidents* trilogy, 2008-2018). Port Orchard, WA: Leafcutter Press.

in the record without any evidence of predecessors.[11] These animals show up highly specialized, which, in evolutionary terms, means we should see evidence of their evolutionary forerunners. The problem is, we *don't* see them. This is true for 99% of the fossils so far discovered. We can't identify who they evolved from much less what they evolved into. Is it because they aren't there at all and so never will be found, or is it due to some mysterious artifacts of the fossil record? This debate has been waging for decades. Unfortunately, this problem is recognized by a good many paleontologists as owing to an "incomplete" or "poor" fossil record, rather than a strong indicator of a falsified theory of evolution. True to the Lewontonian approach, consensus-holding naturalistic evolutionists stick to their "unsubstantiated just-so stories," where they engage in endless erudite speculations about the mysteries of evolutionary rates as well as who the "likely" candidates are in the myriads of differing phylogenetic trees they come up with.[12]

The second overwhelming piece of disconfirming evidence from the fossils is stasis. What we see is that when a critter shows up in the fossil record it remains unchanged throughout the entire time it is present in that record.[13] This is also true for 99% of the fossils we know about. The fact that there are so many species that never changed during their entire appearance in the fossil record (even up to the present day) is one reason why paleontologists have been scratching their heads endlessly ever since Darwin in their attempts to discover the expected countless evolutionary precursors. Instead of seeing such evidence everywhere, as evolutionary theory predicts, it is practically nowhere to be found.[14] And where scientists think they have established evo-

---

11 Carroll, Robert. 1997. *Patterns and Processes of Vertebrate Evolution*. NY: Cambridge University Press, p. 2. "Instead of showing gradual and continuous change through time, the major lineages appear suddenly in the fossil record, already exhibiting many of the features by which their modern representatives are recognized." In actuality, many of the life forms we know of today look *identical* (allowing for normal variations) to their ancient counterparts.
12 Feduccia, Alan. 1999. *The Origin and Evolution of Birds*. New Haven: Yale University Press, p. 59. "cladistics is thus not without criticism, and many of the criticisms are lodged at the notion that cladistics is a panacea for all systematics. ...Halstead has criticized cladistic methodology for the apparent precision and respectability that cladistics confers by its cladograms on what is in reality no more than speculation."
13 Carroll, Robert. 1997. *Patterns and Processes of Vertebrate Evolution*. NY: Cambridge University Press, pp. 3-4. "...It must be assumed that evolution occurs much more rapidly between groups than within groups. For most of their evolutionary history, fundamental aspects of the anatomy and way of life of these lineages do not change significantly. Very few intermediates between groups are known from the fossil record."
14 Feduccia, Alan. 1999. *The Origin and Evolution of Birds*. New Haven: Yale University Press, p. 59. "The question remains, though, what antedated Archaeopteryx? What were the reptilian ancestors from which it and other birds descended?...after much more than a century of investigation and a fairly satisfactory fossil record of reptiles, the answer remains highly controversial." The reason it remains controversial is because Archaeopteryx, like every other vertebrate we know about, shows up suddenly in the

lutionary relationships is typically due only to gigantic stretches of the imagination. Even when they admit that their imagined scenarios are unlikely, they continue to support them. Once again, this is consistent with the Lewontin Imperative.

The point is this: the fossil record does *not* provide confirming evidence of progressive evolution within the alleged vast eons of history. Many reputable scientists who understand fossils, especially paleontologists with a lifetime of effort looking, readily admit this.[15] Since they cannot fabricate fossils, they have to find a way to explain away these two very basic anomalies in the fossil record. This is one reason why the evidence from fossils is perhaps more important than any other type of evidence needed to confirm evolution.

So, how do they explain this lack of evidence which even Darwin himself noted was deadly to his theory? They nuance and massage the evidence using a clearly Lewontonian approach. They begin with the (materialist) assumption that evolution is a fact and then attempt to explain why the evidence fails to demonstrate it. They insist that there must be something "missing" in our understanding of the evidence, thus, the expression of "missing links." This alleged missing evidence is referred to as "gaps" in the fossil record. But the idea of gaps strongly implies that the missing fossils *must have* existed at one time or another. The assumption is the links are out there somewhere. So why aren't they being found? What if there really are no "missing links"? This is a clear indicator from the evidence at hand, but many consensus holding evolutionists are simply unwilling to consider such a notion.

Perhaps the most sobering implication of this entire debacle of the lack of evidence in the fossil record is the implicit meaning underlying the facts. If the fossil evidence fails to demonstrate evolution, what then is the basis for accepting it? The answer is two words: unyielding faith. This faith is described in the Lewontin Imperative as the unconditional acceptance of materialism, *no matter what the evidence.*

Many of the individuals in the case studies of this book were influenced by practitioners and enforcers of the Lewontin Imperative, and the

---

fossil record and with no known parental stock. Even the lowly trilobite, an underwater arthropod found the world over in quantities exceeding perhaps any other known fossilized animal, provides us with the same evidence, only in the case of the trilobite this problem is vastly more exacerbated because there are literally hundreds of different species of trilobites, and we have no compelling progressive evolutionary history telling us where they came from.

15 Gould, Stephen Jay. 2002. *The Structure of Evolutionary Theory.* Cambridge, MA: The Belknap Press of Harvard University Press, pp. 750-755. On these pages Gould recounts the reports of 10 or so workers who report that their lifetime efforts have failed to identify progressive evolution among the critters they studied.

consequences have been massive and far-reaching. In addition to misleading others via expertly nuanced perceptions of reality, what makes the Imperative particularly onerous are the predictable actions by many self-appointed enforcers, as they relentlessly seek to ensure the dictates of the Imperative are carried out through various tactics of intimidation and discrimination.[16] The imperious enforcers of the Lewontin Imperative frequently seek to ruin the careers of those who call evolutionary dogma and alleged discoveries into question. Such dissidents are frequently attacked as being unintelligent and uneducated "science deniers,"[17] when in fact they represent one of the most sacred virtues of scientific inquiry, namely, dissent. Under the influence of the Imperative, the virtue of dissent when it comes to evolution has become twisted into a macabre pseudoscientific, religiously motivated notion that must not be tolerated. Silencing Darwin dissidents has been a key objective of many enforcers within the evolution community for decades, and with perceived wins in court battles behind them, they feel emboldened to continue with their career-ending schemes.[18]

One of the results of purging dissenters is, only those who agree with the materialist evolutionary consensus views remain standing. It is a dishonest characterization to suggest evolutionary theory is gaining wider acceptance (as indicated by polls) by proclaiming more people are finally becoming better educated and therefore are coming to the conclusion that Neo-Darwinian evolution is true. Actually, the growth in acceptance of evolutionary views are due primarily to a key outcome of implementing the Lewontin Imperative: eliminate your opposition. The result is, only those who agree with the Darwinian dogma remain to influence the next generation.

---

16 For example, Jerry Coyne, whose Ph.D. thesis advisor was none other than Richard Lewontin. In late 2018, Coyne went ballistic on a huge rant and harangued Springer publishing house for allowing the publication of an article he labeled as "tripe," and a "travesty." His actions in that incident reveal the results associated with enforcing the Lewontin Imperative. His example is but one of thousands of similar actions. But Coyne's actions also reveal something many others of his ilk are acclimated to. They seem to forget that if you seek to restrict speech you don't like, then it's clear you don't support freedom of speech. And that is part of the crux of the intolerance exhibited by many Lewontin Imperative devotees who see themselves as self-appointed enforcers and censors.
https://whyevolutionistrue.wordpress.com/2018/12/18/creationist-paper-gets-into-a-springer-journal/ https://whyevolutionistrue.wordpress.com/2019/03/06/retraction-watch-highlights-the-paper-i-got-retracted/ https://whyevolutionistrue.wordpress.com/2019/03/01/springer-apparently-retracts-a-creationist-paper-but-its-still-on-the-website/
17 Prothero, Donald R. 2017. *Reality Check: How Science Deniers Threaten Our Future*. Indiana University Press. Prothero and others place Intelligent Design advocates into the same bucket along with flat-earthers and those who believe the moon landing was faked. This "guilt by association" tactic is an ad hominem logical fallacy.
18 Bergman, Jerry and Kevin Wirth (contributor). 2016. *Silencing the Darwin Skeptics: The War Against Theists* (Vol II in the *Slaughter of the Dissidents* trilogy). Leafcutter Press.

Other less obvious tactics combine to promote the significance and acceptance of evolution over dissenting views in ways many of us never even notice. Take, for example, the publication at one web site of the trial transcripts provided for pro-evolution expert witnesses at the Arkansas trial of 1981, but no transcripts of the court testimony are similarly provided from witnesses on the creation side.[19] Providing transcripts for only one side effectively censors the views of the Darwin skeptics who testified, and prevents others from reading about the evidence they presented.

So I ask readers the obvious question: how then can we know what to trust coming from the evolution camp? How can anyone, including the cases of people presented here by Dr. Bergman, make a decision to side with an evolutionary worldview when we see this sort of approach being utilized?

Imagine how different the world might be if a P.Z. Myers, Louis Leakey, or Michael Shermer had not renounced their former Christian faith. Then think about all the people they have influenced away from the notion of a Grand Designer as a result of their lifetime of effort. That sobering thought should serve to galvanize in readers the realization of just how important it is to make sure some helpful lifelines are thrown to anyone struggling with choosing between a designer-oriented worldview and the nuancing and coercion associated with promoting evolutionary arguments and "evidence."

The companion book to this work, *Science is the Doorway to the Creator*,[20] is an excellent contrast and a counterpoint to this volume, as well as a source of encouragement. It presents individuals who effectively engage in the scientific enterprise who also do not support the notion of Darwinism. These well-educated people with keen minds serve as great examples of how people holding to a non-evolutionary worldview can work competently in their scientific field of endeavor. I mention this because so many critics of Darwin dissidents make the false claim that people who believe in a Grand Designer are scientifically incompetent, and are quite often treated as a serious threat to the overall credibility of science.[21]

This book serves as both a warning and a resource for, not only young people in search of answers, but also for parents, science educators, pastors, and other people who find themselves in a position to provide some guidance

---

19 McLean v. Arkansas Documentation Project. http://www.antievolution.org/projects/mclean/new_site/index.htm#Plaintiff's transcript
20 Bergman, Jerry. 2019. *Science is the Doorway to the Creator*. Spokane, WA: Leafcutter Press.
21 Bergman, Jerry with Kevin Wirth (editor and contributor). *Slaughter of the Dissidents*, Leafcutter Press, 2008-2018. See the *Slaughter of the Dissidents* trilogy for over 100 case studies about how creationists and Intelligent Design advocates have been persecuted and discriminated against in recent decades.

and thought-provoking material supporting the idea of a Grand Designer of life. As this book documents, failing to provide the help young people need to navigate through the legitimate questions they have could result in some very dire consequences. A key purpose of this book is to provide a few resources to those who desire to find answers to some of the most challenging questions and claims of evolutionary science (see the appendices at the end of this book). For those who intend to use this book as a means to help others who are working through these important issues related to evolution, I urge you to do whatever you can to ensure those in your sphere of influence find valid answers to their questions and help them come through their inquiries on this subject with their design-oriented worldview intact. The answers to their questions are incredibly important and available, if only someone would point them in the right direction.

Kevin H. Wirth
October, 2019

ENDNOTE: Some of the advance reviewers of this Introduction commented to me that they thought I was attacking or being too harsh on Mr. Lewontin. Many of them suggested I should be thanking him for being so candid about his views. Well, I DO wish to thank him for his openness, but the point is his comments indicate a serious problem among many evolutionists who, like Lewontin, insist on using their materialist worldview to color and frame all of evidence they are privy to. They do this, as Lewontin insists they must, "***in spite of*** *the patent absurdity of some of its constructs,* ***in spite of*** *its failure to fulfill many of its extravagant promises of health and life,* ***in spite of*** *the tolerance of the scientific community for unsubstantiated just-so stories."* My comments about Lewontin and his "Imperative" are not intended to be a personal attack on Lewontin, but to use Lewontin's comments to illustrate the near universal attitude of those who agree with him and operate under the dictums of the Imperative. The point here is that many evolutionists come up with "just so" stories that they try to fit into an evolutionary framework, even when the evidence to support such stories fails abysmally – they fail to see or admit any problem because their backbone is their story-laden theory, and they continue to devise rationalizations, no matter how feeble, that align with their theory rather than admit to other more likely or equally viable possibilities. Creating unsupportable explanations in an effort to sustain the primacy of any existing paradigm is not good science. Such claims provide no reasonable basis for acceptance wherever they appear and deserve to be called out for what they are.

Introduction by Dr. Jerry Bergman

The purpose of this book is basically threefold:

a. Provide readers with information about what has happened to many individuals who were confronted with a crisis of faith as they began to wrestle with Darwinism.

b. Identify and explore some of the views and reasons why many decided in favor of a Darwinian worldview and the impact of that decision.

c. Offer solutions for combatting this challenge by providing access to reasonable evidence and arguments demonstrating why the acceptance of Darwinism and the rejection of a Designed Universe goes against fact and reason.

Many of the individuals reported here had a Christian upbringing or influence in their youth, and yet owing to a number of factors, they converted to a Darwinian worldview once they were confronted with what appeared to be evidence strongly favoring evolutionary concepts. The evidence they ran into in one or more epiphanies or encounters (called "Turning Points" in this book) are instructive to obtain a better understanding about why people are persuaded to discard their creation-oriented perspective in favor of an evolutionary worldview.

Understanding these Turning Points is critical to consider what information is required to help others who similarly wrestle with the same issues. There exists a good deal of scientific evidence that can successfully challenge the Darwinian paradigm. The following patterns of influence are observed in the lives of many persons reported in this book who wrestled with this issue. The impact of these patterns will be summarized at the end of this introduction.

1. *<u>Evolution is a key influence on the road to atheism</u>*.

Many have suspected this to be true, but this book gives substance to this suggestion and moves it into the realm of reality. A key goal of this book is to document this problem. Throughout this book, I only briefly respond to a few of the more egregious claims made by ex-creationists and ex-IDers. A far longer book would be required to respond in detail to all of the claims made by many individuals who have made this philosophical journey.

Why Darwinism is the Doorway to Atheism

The main reason people give for believing in a creator of life is the evidence of the beauty and wonder of creation. Darwin realized this, and knew if he was able to convert the world to his view (and this was clearly his lifelong goal since he was a young adult), he would have to produce another theory of creation. His trip on the *Beagle* proved to be the seminal influence for the development of his ideas. The theory he came up with was the survival-of-the-fittest notion. The more fit life forms would cull the weaker, less-fit, and slowly evolve to become better adapted to their environment. Darwin's idea has become the most contentious notion about the history of life ever since he first published it in 1859.

This study documents the common progression that occurs when someone is challenged to shift their philosophical outlook on life to adopt an evolutionary worldview – and how this shift often results in a slide towards atheism. A significant number of persons who at one time held a Young Earth Creation (YEC) or Intelligent Design worldview have become evolutionists (often very militant), and many have become atheists as well. Over 100 cases of leading evolutionists were explored to understand their reasons for rejecting the creation worldview and accepting Darwinism. Some cases began with problems around young-Earth creation and then progressed to full-fledged evolution. One example is Anthony Lawson Larson "who started out as a young earth creationist, then embraced the ID movement in its early history, and then eventually moved briefly to old earth creationism, and finally to evolution."[22]

The late Wilbert H. Rusch Senior (Professor Emeritus of Biology and Geology, Concordia College, Ann Arbor, Michigan) observed this as part of a small geology graduate seminar in his late forties, with the rest of the group of men in their early twenties. One afternoon he

> was participating in an impromptu rap session on the subject of Christian beliefs. In the course of the discussion I was greatly disturbed to discover that the whole group were apostate Christians

---

22 See https://www.amazon.com/gp/review/R22J1XV6767DN6?ref_=glimp_1rv_cl

of various mainline denominations. As the discussion progressed, it developed that for each of them, their Christian faith had been eroded over a period of time as their acceptance of the theory of macroevolution grew.[23]

Kenneth R. Miller,[24] a biologist at Brown University, was an expert witness in the Kitzmiller v. Dover court case that expelled "intelligent design" from Pennsylvania public school classrooms in 2005. Miller, who claims to be a Christian, concluded about the ruling resulting from this case "We're moving in the right direction" (towards evolution). Miller added he was very encouraged by the increase in younger people embracing evolution and called it "quite striking."

Rachel Gross, a supporter of the evolution worldview, added that, according to a 2014 Gallup poll, it is "not just the young who are moving in favor of secular evolution," by which she means atheism,[25] but the "overall proportion of Americans who believe in secular evolution has doubled since 1999, from 9 percent to 19 percent." This translates into the "nones" being the "fastest growing religious cohort in America."[26]

In some cases, it was not clear if the person labeled a none was an atheist, agnostic or simply a "none" or another label, however it was clear that they were all evolutionists. The typical progression of those who make the philosophical shift from creationism or Intelligent Design is then to adopt theistic evolution, then move to evolution with a thin veneer of theism, to a *none*, and lastly an atheist, but it is not always obvious where someone is on this continuum. Furthermore, many people do not move through all these stages before settling on any particular view. Many of our universities encourage students with concerns about Darwinism to take a step towards a TE worldview. Former atheist Wayne Rossiter, who converted to Christianity as an adult, noted

> The Christian student sitting in a high school or college classroom is told not to be uncomfortable with what Darwin has to say. Our educators point to names like Francis Collins or the late Theodosius Dobzhansky, and say, "see, these scientists are Christians,

---

23 Rusch, 1991.
24 Kenneth R. Miller. (2008) The Collapse of Intelligent Design (Video Lecture) https://www.youtube.com/watch?v=Ohd5uqzlwsU
25 This is the normal outcome from adopting a naturalistic evolutionary viewpoint.
26 Michael Shermer. 2013. p. 82.

and yet they accept Darwin." So the theist is being asked to fully ascribe to Darwinian evolution. But none of these educators, lecturers, or writers are making an equally forceful case to atheistic evolutionists that, "These evolutionists also believe in God."[27]

The Loss of Faith is linked to Turning to Darwinism

A key cause for the loss of Christian faith is often the acceptance of Darwinism. Darwin has been enormously successful in his goal of eradicating the notion that God was responsible for the creation, which was, in his words, "like confessing a murder."[28] The success of this goal is measurable today. Under the headline: *Evolution Is Finally Winning Out Over Creationism*, Rachel Gross wrote:

> While the majority of people in Europe and in many other parts of the world accept evolution, the United States lags behind. Now, at long last, there seems to be hope: National polls show that creationism is beginning to falter, and Americans are finally starting to move in favor of evolution. After decades of legal battles, resistance to science education … evolution may be poised to win out once and for all.[29]

The reason for the shift is due to the change in the beliefs of younger generations. A Pew Research Center report found that fully

> 73 percent of American adults younger than 30 expressed some sort of belief in evolution, a jump from 61 percent in 2009, the first year in which the question was asked. The number who believed in purely secular evolution (not directed by any divine power) jumped from 40 percent to a majority of 51 percent. In other words, if you ask a younger American how humans arose, you're likely to get an answer that has nothing to do with God.[30]

She concluded "most of that increase has been drawn from the pool of Americans who previously reported that they believed in evolution guided by God, which simultaneously dropped from 40 percent to 31 percent." In oth-

---

27 Rossiter, 2015, p.6.
28 Darwin, Charles. 1844. *Letter to Joseph Hooker.* January 11.
29 Gross. 2015.
30 Gross. 2015.

er words, theistic evolution is often a doorway to atheism. Furthermore, she added there are several signs that these numbers reflect a shifting cultural tide:

> First, America is getting less religious. Today's younger Americans no longer have the strong ties to organized religion that their parents did. About 56 million people now call themselves "nones" —meaning that they identify as atheist, agnostic, or nothing in particular on national surveys—a jump of 19 million since 2007, according to the 2014 Pew Research Center survey. Again, it's the younger generation who are driving this shift: Fully 36 percent of young adults between 18 and 24 identify as nones, and the number of millennial adults who are religiously unaffiliated is growing fast.[31]

She added "the fact that fewer people are identifying with an organized religion is good news for science education, because many of those [conservative] religions have historically opposed evolution." Furthermore, the persons who still believe

> that evolution is a myth and that humans have existed basically as-is from when they were created are older Americans. About 34 percent of Americans 50 to 64 years old believe in creationism. For Americans older than 65, it's 37 percent. From the perspective of people who endorse evolution, that's a good thing—because, not to be insensitive, but old people die. When these elderly creationists shed their mortal coil, they will be replaced by that younger generation consisting increasingly of nones. The result: a steady phasing out of those who oppose evolution.[32]

2. *Many individuals began their lives with a Christian upbringing but discarded it when they encountered Darwinism*

For the last 40 years, I have read an average of one book a week, mostly on genetics, evolution, creation, Intelligent Design, and science in general. Of these, I noticed over and over that many of the authors, mostly scientists, were reared as creationists and became evolutionists and, not uncommonly, atheists as well. I selected the cases that I felt best illustrate the theses of this

---

31 Gross, 2015.
32 Gross, 2015.

book, a difficult task considering the number of cases that I had available. Most, but not all, of the cases that are documented in this book are about Ph.D. level scientists or college science professors. Every person was male except two, which may be another area to explore.

I have also found that a surprising number of the last century's leading evolutionary biologists were reared as creationists, or at the least, as conservative Christians, and accepted creationism as youths but rejected it as adults. This included not only Charles Darwin, but also many of those discussed in this book. Professor Chad Walsh wrote that:

> Fundamentalism invites a conflict between religion and science (many of the most confirmed atheists I have known were reared as sturdy Fundamentalists and then studied geology or biology); Modernism resolves the supposed conflict by *making religion so vague that it won't get in the way*.[33]

Some people who post on the internet about their movement from Christianity to atheism as a result of their encounter with Darwinism prefer to remain anonymous, such as the following:

> I'm an atheist. I grew up in an evangelical Protestant household. I went to church most Sundays and even went to a Christian school from sixth to twelfth grade. I was a sincere believer from the time I was old enough to understand Christianity until the age of about twenty, although my "official" conversion occurred at age seven. I was also a young-earth creationist for most of that time.

> It's hard to pinpoint the exact time that I started having doubts, but they probably began in 2007, toward the end of my senior year of high school. I fought off these uncertainties for a few years, but in 2010 I began investigating my faith in earnest. After reading several books and countless articles, I left Christianity and eventually became an atheist. I started this blog at the beginning of 2011 in order to explicitly lay out the reasons I no longer believe. Recently I graduated from UCSD with a bachelor's degree in cognitive science.[34]

---

33 Walsh, 1949, p. 74.
34 Anonymous. *Reflections from the Other Side: Leaving Christianity and Embracing Skepticism* (blog) This blog site represents the sentiments of millions of young people in our culture today who drift away from Christianity and embrace an evolutionary world view. For example, atheist Jerry Coyne is one of his

This is a fair representation of the struggle and experience for many who find themselves confronted with Darwinian evolution. In many such cases, no one was able to effectively intervene and provide them with reasonable evidence for rejecting an evolutionary perspective.

3. The impact of presenting evolution as a completely reliable scientific explanation.

Evolution is highly revered in our colleges and the mass media as self-evident and unquestionably true, and is presented as a pillar of modern biology with the backing of countless science professionals. Nearly all of the persons researched here became convinced that Darwinism had overwhelming scientific support and, as a result, rejected any creation or design ideology, and quite often eventually Christianity. My research indicates that no small number of persons fit this scenario, including even many in-the-closet evolution professors at Christian colleges. Some claim to be theistic evolutionists, which usually means they advocate molecules-to-man evolution with a thin veneer of mystical theism.

Ironically, one of the most important scientific discoveries that has helped creationists is one made by Mary Higby Schweitzer, a Christian and paleontologist at North Carolina State University. Science magazine wrote

> After earning an undergraduate degree in audiology, Schweitzer married and had three children. She went back to school at Montana State University in Bozeman for an education degree, planning to become a high school science teacher. But then she sat in on a dinosaur lecture given by Jack Horner, now retired from the university, who was the model for the paleontologist in the original Jurassic Park movie. After the talk, Schweitzer went up to Horner to ask whether she could audit his class.
>
> "Hi Jack, I'm Mary," Schweitzer recalls telling him. "I'm a young Earth creationist. I'm going to show you that you are wrong about evolution."
>
> "Hi Mary, I'm Jack. I'm an atheist," he told her. Then he agreed to

heroes.
http://othersidereflections.blogspot.com/p/about-me.html

let her sit in on the course.

Over the next 6 months, Horner opened Schweitzer's eyes to the overwhelming evidence supporting evolution and Earth's antiquity. "He didn't try to convince me," Schweitzer says. "He just laid out the evidence."

She rejected many fundamentalist views, a painful conversion. "It cost me a lot: my friends, my church, my husband." But it didn't destroy her faith. She felt that she saw God's handiwork in setting evolution in motion. "It made God bigger," she says.[35]

Although her discovery has been massaged into an evolution-oriented explanation, other workers, like microscopist Dr. Mark Armitage, have made similar discoveries but concluded that the soft tissue he found in a Triceratops horn he discovered could not possibly have survived for millions of years.[36] The implications of more soft tissue finds could potentially pose a significant impact on many evolutionary assumptions.

4. *There were key 'Turning Points' which persuaded people to drop their Christian faith for evolution.*

Most of the people in these case studies experienced one or more Turning Points (what they saw as compelling pieces of irrefutable scientific evidence) when they began exploring evolution in depth. The experience for each person was different with respect to which Turning Point(s) had the most impact. Many were completely unprepared for what they found in evolutionary arguments, and were not able to effectively overcome the claims presented by evolution. Part of the purpose of this book is to present readers with resources that address some of the more common arguments that turn people away from their ID or creation worldview. The companion volume to this book, *Science is the Doorway to Theism*,[37] also responds to many of the arguments put forward in this volume, thus only brief comments are included in response to the anti-creation and anti-ID arguments of those documented

---

35 Service, 2017, pp. 1088-1091.
36 Armitage, Mark. 2016. "Preservation of Triceratops horridus Tissue Cells from the Hell Creek Formation, MT." *Microscopy Today*, January. pp: 16-22.
_____, and Kevin Anderson. 2013. "Soft Sheets of fibrillar bone from a fossil of the supraorbital horn of the dinosaur Triceratops horridus." Acta Histochemica 115 603–608.
37 Bergman, 2019.

in this volume.

5. *The impact of Theistic Evolution (TE)*

For many individuals who accept the credibility of both a design-oriented worldview and the evidence of evolution, a compromise solution seems to be the best answer. TE is seen by many as that compromise[38] Yet there are many instances where those who adopt this position later move on to agnosticism or even atheism. Ultimately, the eventual conclusion is that God is not needed because evolution explains everything.

This is evidenced time and again as I read countless stories, including many of the individuals covered in this book, who as young people sought to reconcile their Christian faith with what was being presented to them as reliable evolutionary science. One young man published his story of progressing from a Young Earth creation apologist to a TE after attending a seminary. After reading a couple of books, he began to reconcile his growing questions about the inconsistency between evolution and God as a creator.

> When we speak of the evolutionary process, we are speaking of the cause by which new species arise. When we speak of God as creator, we are speaking about the agency behind the cause. In other words, science gives us the mechanism by which life develops, and religion gives us the agency behind the mechanism. Science and religion are not in opposition to each other, but complement each other.

This is a common reasoning behind why many TE's settle on their viewpoint. He continued his account of how, and why, he adopted TE.

> Just recently a friend of mine asked me how I could believe in evolution. "What about the lack of evidence in the fossil record? What about the lack of evidence from genetic mutations? Evolution has no answers for this." Of course, now I know that there are indeed good answers for these questions, based on solid scientific research. But I know from my own background that these aren't the "answers" he really wanted. His point was that because the theory of evolution cannot provide all the "reasonable" answers (in

---

38 For a good review of the major problems with a Theistic Evolution worldview see *Shadow of Oz* by Wayne D. Rossiter (2015), Eugene, OR: Pickwick Publications. Dr. Rossiter is a Professor of Biology at Waynesburg University.

his mind) to *every possible question*, it cannot be true.³⁹

Here he reveals not only perhaps an erroneous judgmental view of his friend's thinking (ie, "not the "answers" he really wanted…"), but he made the same mistake about evolutionary 'evidence' that many individuals make when they rely on the opinions of evolutionary experts, who are obviously biased to nuance and promote an evolutionary story at almost every turn. In fact, the fossil record is replete with a powerful and ubiquitous message of *stasis* (non-change) that clearly shows us that fossils do not tell us any story about evolution at all. But we would only know this by reading books on vertebrate paleontology and seeing firsthand how the explanations are all presented in terms such as "must have," "is assumed to be," "most likely evolved," "is thought to be indicative of," "no intermediates are known," and so on. So the reality is, there is actually no good evidence for evolution in the fossil record, according to paleontologists who know this beyond a shadow of a doubt. But they push the fossil evidence in an evolutionary context anyway because they are convinced about the reliability of the theory even when the evidence fails to support it. This means that the basis for this young man's decision to incorporate an evolutionary perspective into his TE worldview is distorted and defective.

One additional big problem with TE, from a theological perspective, is that it typically denies the reality of key events and individuals recorded in the Bible, especially Genesis. Adam and Eve, Noah and the flood, and other matters are often seen as symbolic rather than literal, even though Jesus is recorded in the New Testament as speaking about them as real people and events. As one contributor wrote in the atheist oriented magazine *Free Inquiry*

> If there was no Adam, there was no fall; and if there was no fall, there was no hell; and if there was no hell, there was no need of Jesus as the Second Adam and Incarnate Savior, crucified and risen. As a result, the whole Biblical system of salvation collapses.⁴⁰

The move from taking the biblical account at face value often results in a slide where other events, destinations, and conditions mentioned in the biblical record are also viewed as symbolic, such as the creation of life, hell, and the concept of original sin. Taking the position of TE results in some

---

39 Russo, Mario. 2015. "Addict: From Young-Earth Apologist to Evolutionary Creationist." *Biologos*, December 7.
https://biologos.org/blogs/brad-kramer-the-evolving-evangelical/tales-of-a-recovering-answer-addict-from-young-earth-apologist-to-evolutionary-creationist
40 A.G. Mattill, Jr. "Three Cheers for the Creationists," *Free Inquiry*, vol. 2, Spring 1982, p. 17

significant theological contortions, which writer Thomas Freeman has summarized.

> Since [*in the mind of a Theistic Evolutionist*] the Creation and the fall are mythological and not literal accounts there is no basis to assume there is original sin. The gospel of Jesus Christ is based upon original sin. The death of Jesus upon the cross is a sacrifice (propitiation) for that sin. Paul argues it, "As in Adam all have sinned and died so in Christ all are made alive." Since there was no Adam, Eve, Eden, or Fall then there is no basis for a need of salvation and hence no basis for Christianity to even exist in our modern era.
>
> Theistic Evolution is a denial of the Gospel. It was the physical death of Jesus that paid for the sin that led to Adam's physical death (and that which brought death into the world). Therefore those who embrace theistic evolution are embracing an untenable position which is contrary to the gospel.[41]

TE also results in a worldview shift that makes the idea of evolution seem much more plausible. After all, if the biblical creation account isn't to be understood literally, then God could certainly have used the process of evolution to create life. Makes perfect sense to the person who has discarded a historical reading of the biblical account. This is a pattern we see in the lives of many individuals reported here. But even more important is the fact that the slide into TE often leads many to eventually embrace a growing skepticism about many biblical passages, which leads to agnosticism and even atheism.[42] This surprising shift occurs in the lives of people who many Christians would think would never go there, like Billy Graham song leader Charles Templeton, Humanist Chaplain Bart Campolo and George Perdikis, co-founder of the Christian band known as the Newsboys (there is a chapter devoted to each of them in this book).

The potential for such a precipitous worldview shift is the direct result of dismissing ID or compromising the biblical account when the idea of evolution seems to be much more compelling and reliable than what the scriptures say. The denial of the fall eliminates the possibility of original sin and there-

---

41 Freeman, Thomas. 2014. "Theistic evolutionists are one step from atheism." *News 24*, December 9. Emphasis added.
https://www.news24.com/MyNews24/Theistic-evolutionists-are-one-step-from-atheism-20141209
42 Hailes, 2017.

fore removes any need for Jesus to have saved us from anything. And if that is the case, then there is no need for Christianity, and atheism, which is right at home with evolution (according to PZ Myers and others), seems to be a much more realistic proposition.

6. *Effective counterarguments to evolutionary propositions are not widely known*.

Lack of awareness of the many scientific, logical, and philosophical challenges to the Darwinian paradigm remain obscure or unknown to many who struggle with this issue. This is primarily because most academic institutions are reluctant to point out problems with the theory for a variety of self-preservation reasons. In reality, there are many sources of information available for questioning the evidence offered in support of evolution, and some excellent resources are presented in the appendices at the end of this book.

Finally, another factor for why problems with evolution are not very well known is primarily owing to the fact that most universities fail to present compelling counter-evidence to evolutionary assumptions and claims. Publisher Kevin Wirth recalls that it wasn't until he was completing his sophomore year in college that he had ever encountered any evidence that challenged conventional evolutionary propositions, noting

> All along the way in my education up until I was a sophomore in college, no one ever even hinted to me that there were any problems with evolution. I was dumbfounded when I began doing my own research on this subject and discovered that many intelligent and accomplished scientists actually had remarkable evidence and arguments that called into question many of the beliefs I had until then absorbed and accepted based on the authority of science experts. My Theistic evolutionary views quickly developed many holes which eventually resulted in the sinking of my evolution perspective and provided me with a newfound rationale for a creation worldview.[43]

It's quite common for many young people entering college to encounter a strong pro-evolution lobby which presents Darwinism as an unassailable certainty, without any problems being identified as significant or problematic.

---

43 Conversation with Kevin Wirth

The Role of Mutations in Darwinism

One of the most common and easily refuted pieces of evidence presented in support of evolution is the notion that mutations play a key role in the evolutionary modification of organisms, enabling them to be better adapted to their environments. After Darwin, the idea that genetic variations could be supplied by mistakes (mutations) in the DNA was combined with Darwin's theory of natural selection. This view became known as Neo-Darwinism. The goal of Darwinism has been enormously successful in spite of the fact that close to 99 percent of all mutations are either near neutral or harmful. Near neutral means that each individual mutation has a very minor effect, or in a few cases possibly none at all, but collectively they add up, eventually producing genetic meltdown, often called genetic catastrophe. What this means, essentially, is that evolution continues to hobble along as the most important pillar of biology whilst being seriously crippled using this impossible claim. Natural selection as a tautology doesn't fare much better.[44] Of course, evolutionists claim that eons of time would allow such mutations to work their evolutionary magic, but this is speculation, not evidence.

Near neutral mutations often cause aging as well as many diseases, including cancer. In fact, we know that the number of genetic mutations in life is cumulative, and unless genetic intervention occurs, it will eventually cause the extinction of all higher life.

An estimate of the mutation rate is achieved by using genetic analysis to measure the family's genetic pedigree and comparing this with the parents offspring. A comparison can then be made to determine the number of new mutations. The new harmful germline mutations that are not repaired or removed are passed on to subsequent generations, causing a gradual accumulation of the number of mutations. The numerous studies that have measured this rate in humans found it to be between 75 and 175 new mutations per generation.

A variety of researchers have used this mutation rate data to conduct complete computer simulations to model the results of the accumulation of mutations in the human genome. This incessant process of genome degradation with each successive generation is an example of what is known as genetic entropy.

If the mutations are harmful to the degree that they cause early death before an organism is capable of reproduction, those genes will not be passed

---

44 Brady, 1979.

on to future offspring. Consequently, in this case these mutations would theoretically be eliminated from the gene pool. This buildup of mutations eventually would allegedly reach a critical level; thus, humans would eventually go extinct at the point called mutation meltdown or error catastrophe. In this scenario, evolution would be going the wrong way.

7. *The influence of mentors and thought leaders*.

Many individuals I reviewed found themselves impressed by one or more evolutionary thought leaders who expressed what seemed to be compelling arguments favoring an evolutionary worldview. Many read books, attended lectures, or watched videos where the views of evolutionary speakers were authoritatively articulated. Modern day thought leaders promoting evolution and typically atheism are not difficult to find. Some of them include Carl Sagan, Richard Dawkins, Bill Nye, Stephen Hawking, Jerry Coyne, Francis Crick and James Watson, Richard Leakey, and P.Z. Meyers. Individuals contemplating how they will resolve their position on this issue would do well to not just listen to the silver tongued pied pipers of evolution, but analyze the subjective nature of their arguments, ask questions about the actual evidence, and not rely on projections, conjectures, extrapolations, assumptions and the like.

8. *The critical influences in academia*.

Dr. Cameron Wybrow wrote that many professors and pastors "who are now in senior positions at various liberal Christian colleges and liberal Christians churches, where their side has triumphed… grew up in YEC homes or at least YEC churches when YEC was in its heyday under Gish and Morris." Furthermore, "the fight of their life was for Christians to be allowed to believe in evolution [and] they are not about to put the achievement of their lives into question by being open to criticism…most of the hard line theistic evolutionists (TE's) are likely to be in that age group."[45]

Another trend noted in this study was that problems with a YEC worldview quite often arose over considerations of the age of the Earth and the universal Flood. Once this worldview is seriously questioned, the YEC supporter may become a long-age creationist or, more commonly, a theistic

---

45   Letter to Jerry Bergman from Dr. Cameron Wybrow, June 2, 2010.

evolutionist,[46] and some even progress to become agnostics or atheists.[47] Thus, for some the age issue may be their turning point to the doorway to atheism.

Rabbi Sacks wrote of his university experience at both Oxford and Cambridge in the 1960s, noting then that "the words 'religion' and 'philosophy' went together like crickets and thunderstorms."[48] Although one often found them together

> the latter generally put an end to the former. Philosophers were atheists, or at least agnostics. That, then, was the default option, and at the time I did not know any exceptions. The first thing we did, a kind of nursery-slope exercise, was to refute all the classic proofs for the existence of God. Kant had disproved the ontological argument. Hume had shown that for any supposed miracle, the evidence that it had not happened was always greater than the evidence that it had. Darwin had shown the error in the 'argument from design.'[49]

Leaving the faith of one's youth is very common among Christians. One study found the percent of those who *no longer* claim to be born-again Christians after four years of college is 27% of those who attended public universities, 45% of those who attended private universities, 31% of those who attended Protestant Colleges and 59% of those who attended Catholic Colleges.[50] After rejecting the creation worldview, these persons typically move into another worldview, which is often, if not usually, Darwinian:

> It's hard to overestimate the appeal of rebelling against the system to a teenaged boy, and that day marked the beginning of my path to a career in evolutionary biology. We learned other things in science class that year, … for example, that all actions have an opposite reaction. For at least one sulky teenager in the small town of Owen Sound, Ontario, it took a creationist to make him into an evolutionary biologist.[51]

---

46 For example, read about the history of changes in thinking about this issue from Dennis Venema. His journey in struggling with the evolution/faith issue is not uncommon. https://biologos.org/files/modules/venema_id_to_biologos.pdf
47 Freeman, Thomas. 2014. "Theistic evolutionists are one step from atheism." *News24*, December 9.https://www.news24.com/MyNews24/Theistic-evolutionists-are-one-step-from-atheism-20141209
48 Sacks. 2011, p. 78.
49 Sacks, p. 78.
50 Wheaton. 2005, p. 174.
51 Canadian Institute for Advanced Research, Botany Department, University of British Columbia,

Another reason for conversions to Darwinism, as stated by one creationist, is that:

> Evolutionists have good reason to crow about their victories in public schools and in institutes of higher learning. Over the years, they have mopped the floor up with creationists, having won virtually every major contest. With creation science having such a dismal record, I think it's time [for us creationists] to look for a better strategy.[52]

Efforts to target regions of the country where a creation worldview is held by a majority of the population are thinly veiled attempts to provide insight into areas where "better" education is most needed. People of faith are described as being "hampered" by their religious views as they progress through college. One such study found that

> ...negative relationship between education and religiosity and a positive relationship between education and acceptance of evolution, but how this manifests in college students who differ in degree of religiosity and prior educational experiences is unclear. We focused our study on the relative importance of education and religion on evolution understanding for college students at a large, public university in the Deep South.

The summary of this report concluded that

> Religiosity, rather than education, best explains views on evolution. In areas of the country where the vast majority of residents believe in God and the literal truth of the Bible, students may be hampered as they enter and progress through college. These same states tend to have lower state science standards and lower levels of educational attainment.[53]

An example of the pressure to conform to Darwinism is illustrated by a "Christian" high school biology teacher Stan Roth, an elder in his Presbyterian church. When student Anna Harvey asked about creationism in his

---

Vancouver, BC V6T 1Z4, Canada. Patrick J. Keeling E-mail: pkeeling@interchange.ubc.ca.
52 Strandberg, 2010.
53 Rissler, 2014.

class, the teacher responded as follows: "When are you going to stop believing that crap your parents teach you?," adding that the student was a "snot-nosed twit" who believed in "nonscientific crap" that had no place in the science classroom.[54]

Another example, although somewhat extreme, was the biology college professor who asked his class the following question: "if anyone here denies evolution, could you please raise your hand?" After several students did so, he told those with their hand raised to stand up. After they stood up he then said: "class, take a good look at the students standing. They are not only ignorant, but the European Human Rights Agency has declared them dangerous." He then demanded that these students all leave his classroom and told them they will not be allowed to return, adding, "You are not welcome in my classroom!"[55]

In fact, many young people today are often more closed-minded, partly because they are being openly indoctrinated into Darwinism. Gross claims that "the anti-evolution, …anti–climate change thinking is… an ideology. It's a refusal to engage with reality." She added that a solution for this is "improving science education," meaning schools should spend much more class time on Darwinism indoctrination.[56]

Furthermore, Gross notes that "evolution is in the cultural air we breathe." The blockbuster movie "Jurassic World" takes as its premise the evolutionary idea that birds evolved from dinosaurs. … the television show The Big Bang Theory—which elevates science, including evolution, far above religion. In one memorable episode of the show, Sheldon, proclaims he plans to "spend the rest of my life here in Texas—trying to teach evolution to creationists." [57]

Gross adds "the message is clear: The fact-ness of evolution, at least to viewers of the show, is indisputable, and creationism is little more than a joke." Gross admits that "evolution hasn't won yet…today, plenty of powerful people are still promoting creationist nonsense, notably Louisiana Gov. Bobby Jindal…For those who attend church or synagogue at least weekly, that number is closer to 50 percent. For white evangelicals, it's 60 percent." If the churches would present the creationist view more effectively, they could do much to stem the cultural decline of a creationist perspective.

Part of the problem is "case law has been on the side of evolution and

---

54 Quoted in Institute for Creation Research. 1999. *Impact,* December, p. 6.
55 Related to me by a student in my college level biochemistry class.
56 Gross, 2015.
57 Ibid.

against creationism for decades," not only approving indoctrination into Darwinism, but pro-evolution court decisions are a major factor in forcing this indoctrination on the young.[58]

The effect of constant exposure to evolutionary theory, bombardment of evolution-only propositions, and the pressure to conform creates an atmosphere that eventually begins to take its toll on even those who would not otherwise succumb to those influences if exposure were limited to moderate doses. Under such all-encompassing pressures, many students suffer from what seems similar to a Stockholm-like syndrome where sympathy for an evolutionary point of view is easily cultivated.[59]

The Case of Raymond Damadian

An example of the pressure faced by students and faculty alike is the experience of Dr. Raymond Damadian, inventor of the MRI technology who, as a faculty member, stated that as a result of his being

> constantly exposed to scientific naturalism (and evolution), I became virally infected with such thinking, eventually reaching the conclusion that there was no God, and thus no longer any practical need for Him in my life. I have since observed that this experience is repeated by tens of thousands of churched young people today after a few semesters at college or graduate school. Looking back, it's now easy to see how I was dissuaded away from faith to embrace the "rock solid conclusions" of science. This pressure to cast off all vestiges of faith was greatest in my life when I became a faculty member at Downstate.[60]

He added that the focus at such schools is on "intellectual aptitude" which amounts to conformity of the cultural trends. Furthermore, in order "to achieve status, recognition, respect, and most of all, tenure" which was a goal of every faculty member, one must conform to the latest scientific consensus. The fact is, as relayed by Damadian,

> the undercurrent powering this world of academia was, as you

---

58 Luskin, 2009.
59 It's not much of a stretch to suggest that students often become so inundated with evolutionary propaganda that they become, in essence, intellectual captives of those who insist the theory is respectable and so well founded that opposition to the idea is not only foolish, but career ending.
60 Kinley, 2015, p. 36.

> might guess, atheistic Darwinian evolution. As such, there was exactly zero room for a "Supreme Being" in the equation. And so, wading into the river, I was carried downstream with the current. One day, I shared my new conclusion about God's non-existence with my tech, who immediately became incredulous, replying that she couldn't believe I had said such a thing. "You know it's a fact," I confidently asserted, unaware that I had become yet another victim of evolutionary fiction.[61]

Damadian then added that, as a result, he had effectively reversed his life, turning his back and mind on his spiritual heritage, which he attributed to the secular academia environment:

> I abandoned my upbringing, as well as the decision for Christ I had made ... in Madison Square Garden. ...I had been an active and dedicated churchgoer throughout my youth, even to the point of being selected vice president of the Pilgrim Fellowship of my church. And the salvation decision I made had deeply impacted me. However, during my time in medical school, strong atheistic currents of the scientific community caused me to drift far from the shore of God's reality and truth.[62]

Sometime later Dr. Damadian returned to his spiritual roots.[63] The selection of cases for inclusion in the book covers only a few of the most well-known ex-creationist cases, all of whom, like Damadian, turned their backs and minds on their spiritual heritages and, as far as can be determined, never looked back. A few claimed to be theistic evolutionists, but often that translates to the belief that evolution created everything, with God somehow mysteriously behind it all.

9. *The impact of evolution-oriented activist groups*.

Many activist groups seek to provide training and education to schools and other government agencies in an attempt to support the Darwin-only approach in public school classrooms.[64] The impact of these groups can-

---

61 Kinley, 2015, p. 36.
62 Kinley, 2015, pp. 36-37.
63 Bergman and Wirth. 2012b. See chapter 12
64 Among them are the *National Association of Biology Teachers* (NABT), the *Freedom from Religion Foundation* (FRF), the *American Civil Liberties Union* (ACLU), *The National Center for Science Educa-*

not be overstated in terms of the influence they have in defining the type of evolution-based information and resources they offer. Their education programs are typically designed to assist teachers who engage in compulsory science education programs in different states. Many of those programs do not expose the serious problems with evolution or consider opposing viewpoints to be valid.

10. *Impact of Peer Pressure to adopt a Darwinian worldview*.

The change in the worldview of many cases reported here affected many other life values. In a survey conducted by the Australian National University, it was found that persons who "believe in evolution" are more likely to support premarital sex and unfettered abortion than those who reject Darwin's theory. In addition, persons who accept Darwinian ideas were "especially tolerant" of abortion. In identifying the primary factors determining these attitude differences, the research report author, Dr. Jonathan Kelley, added that "The single most important influence after church attendance is [the acceptance of] the theory of evolution."[65] Other studies show these results are not unique to the university, nor to Australia, but is a world-wide phenomenon.

This background helps to explain much of what I found in my research. A major reason for the current widespread acceptance of evolution includes social influence and pressure, not scientific evidence—the secular scientific community generally *assumes* that Darwinism is factually true, and this worldview saturates both academic science and popular science publications as well as public presentations by scientists and commentators.

Other examples of the problems that can result from the acceptance of Darwinism include cases where becoming an evolutionist produces family conflicts. An example is the concern that the objections of YEC family members may cause the Darwinist to hide their true feelings in this area to avoid familial confrontations.

11. *The church has failed to effectively address evolutionary arguments and many actually support evolutionary views*.

---

tion (NCSE), the *National Education Association* (NEA), to name some of the major groups. See Kevin Wirth's intro for more information under the heading of Activist Education Organizations
65   *The Australian*, February 1, 2000, p. 6.

Another inescapable observation made from studying these cases is that many of the reasons people reject a Creation or Designer worldview can, and should have been, answered by the church. This book is intended to stress the dire need for effective apologetics programs if the church sincerely would like to stem the current hemorrhaging of its members. For many of the individuals cited in this work, there was usually a key Turning Point consisting of so-called scientific evidence that convinced them to turn away from their Christian faith. In many cases there was also a corresponding lack of effective influence from members of the Christian community to help individuals as they struggled with this issue. Still others, as in the case of Stephen Hawking and even Charles Darwin, had spouses whose influence and stand for their faith was dismissed by their husbands.

One other reason why creation science has such a dismal retention record is because many churches often ignore the issue, or even are on the side of the evolutionists, usually because they assume the Ph.D. "experts" are correct, so why even attempt to challenge them? They reason that, "as a minister, I am no scientist, thus have to rely on them for the truth of this issue." This sad lament is also not true. It may take some effort, but today the internet hosts hundreds of videos and papers challenging evolutionary precepts (see the appendices at the end of this book).

A survey by the Christian Research Association found that, after boring church services (42%), the second most common reason for not attending church is disagreement with the church's teaching (rounded to 35%)—with many citing disagreements with the inerrancy of the Bible and high morality teachings. Disagreement with moral teachings of the church ranked third (rounded to 35%), with disagreements mainly on the churches stand on abortion and sexuality. Negative experiences with Christian people accounted for only 14%.[66]

The two topics that are the most difficult for those who struggle with this issue are the age of the Earth and the universal Flood. Although in the last few decades YEC's have made much progress documenting their views in these areas, their scientific research in areas such as the problem of mutations, the fossil record, the origin of biological information and vestigial organs need to be stressed more than they are now.

---

66 Christian Research Association, 2000. p. 6.

Creation is Seen as a Distraction

Many church leaders don't see the idea of creation as a central tenet of Christian doctrine or belief, and many believe that the contentious nature of the issue causes more heartburn than value, and is a distraction from the main message of the Christian gospel. Strandberg provides what he feels is one good reason to ignore the whole evolution issue, namely because

> salvation is the most important issue for us to be spending our time and energy on, all Christian endeavors need to be productive in the area of winning people for the Kingdom of God. When it comes to soul winning, arguing about creationism simply does not carry any weight. Because of the combative nature of this conflict, the salvation message always seems to be lost in the struggle. I agree 100 percent with my creationist brethren that our ancestors did not swing from trees by their tails. I just don't think we should be conducting a propaganda war to win new converts.[67]

The problem with this unfortunate view, as this book and many research studies have documented, is that evolution is often the doorway to the loss of Christianity, and often of theism as well. The solution is clear: more and better education of the fact that Darwinism has been falsified by science is needed. And since this issue is one that many young people eventually grapple with, we need to meet them there with reasonable arguments and not view it as merely a "propaganda war."

Preaching Darwinism in Church and Seminaries

One Ph.D. who has studied in several university departments of religion concluded that:

> 1. The people who hate religion, or condescend to it, in religious studies departments are most often people brought up in religious backgrounds who have become unbelievers, and are striking out at their former beliefs. (This is quite common in philosophy departments as well.)

---

67 Strandberg, Todd. n.d. "Evolution vs Creation: A Pointless Debate." *Rapture Ready* (blog). http://www.raptureready1.com/rr-ec-debate.html

2. There are also many people in religious studies departments who want to throw out large chunks of their former beliefs, but can't quite bring themselves to make the break and … wish to remain within shouting distance of religion, but don't wish to clearly affirm a specific belief, and a secular religion department allows them not to affirm anything, because their job doesn't depend on any statement of faith.

3. In most cases the second group does more damage than the first because the outspoken atheists and religion-haters (Dawkins, Avalos, Bart Ehrman, etc.) are often irrational, and therefore undergraduate and graduate students can assess their arguments accordingly. Liberals with contempt for traditional religion are often harder to spot because they speak the language of traditional religion, and if naïve conservative students speak too forthrightly about what they believe, they may end any chance they may have of an academic career. I didn't realize what bridges I was burning when I criticized historical-critical methods of interpreting the Bible to Baptist seminary professors, or praised Etienne Gilson and medieval theology to explicitly Catholic (and ordained!) religious studies professors.

4. I naively assumed that all Protestant seminary professors would be glad to hear a student defend holistic readings of the Bible. Little did I know (until life made me wise) that Protestant seminary professors were among the greatest boosters of historical-critical study and if you are a student in a secular religion department, or even in most Protestant or Catholic seminaries you should speak guardedly to all professors and write cautious, "safe" academic essays until you know where your professors are coming from.

5. Liberal seminaries can be as bad or worse than secular religion departments. They are filled with people who fought a long battle to be allowed to endorse evolution. Question evolution, question historical-critical studies, question any of the left-liberal social causes that these professors have embraced, and you will be subtly or unsubtly punished. Many

students have left Protestant seminaries crushed and disillusioned.

6. I was fortunate to have gone through a religious study department in which the atheists and agnostics kept their anti-faith, or lack of faith, out of their teaching, and treated religious students equally with non-religious students, as long as they did good graduate and undergraduate work. It is hard to achieve a balance like that, and I suspect that it has [often] not been achieved. I suspect, however, that this historical moment has passed, and that the theological and cultural left now has such a stranglehold on religious studies departments that it would be hard for a student today to have an experience as I did.

I would advise anyone studying religious studies, divinity, philosophy, English literature, etc. to enter the academy with eyes wide open, and to realize that the general ethos is hostile to people who would even consider the idea that there might be a thing as proven "truth." There are still a few good people generally scorned by their colleagues, who have somehow managed to get tenured in the face of the anti-religious, anti-traditional Thought Police. Find these people. And you're as likely to find them in a history department or a political department, or a Classics department as in a religion department or a seminary.[68]

Many clergy were one of the first, and most active, defenders of both Darwinism and eugenics.[69] Ironically, many ministers of mainline churches, such as some leading Methodist clergy, naively and actively defend Darwinism today. Many of these clergy teach that Genesis is mere poetry or stories that teach some moral precept. Some even actively oppose both creationism and Intelligent Design.

Church of Christ minister Rev. Rich Smith in a recent sermon preached it is "astonishing" that most people today do not accept Darwinism "eight decades after the Scopes Trial." As evidence, he cites Gallup polls

---

68 Email from Cameron Wybrow.
69 Livingstone, 1987.

that show "some 44% of Americans hold to a creationist view," and "another 40% are sympathetic to some sort of Intelligent Design hypothesis."[70] After condemning creationism and Intelligent Design, Rev. Smith then condemned theistic evolution—evidently only functional atheism is acceptable to him. In his words, Intelligent Design is

> just the latest manifestation of "creationism." Of course, on its surface it differs from the old fashioned literal interpretation of scripture. ID allows in fact that an evolution-like process may play a role in creation, but posits that it is purpose driven, that we are not here by chance. Specifically, it says, there are some things in the world, most notably life itself, that cannot be accounted for by known natural causes, and show features that, in any other context, we would attribute to intelligence.[71]

Rev. Smith concluded the idea that "Living organisms are too complex to be explained by any natural—or mindless—process" and "can be accounted for only by invoking a designer—a very, very smart one!" is totally wrong.[72] He argues that this view is a non-starter because the conclusion that God is the "creator of the heavens and the earth" is not "good science."[73] In other words, living organisms are *not* too complex to be explained by Darwinism, and could easily have been created by the natural, mindless, purposeless evolutionary process of natural selection involving genetic mutations combined with other evolutionary mechanisms! In his words, ID advocates who claim that Shakespeare's "work could never have come about by pure chance" are wrong. To prove this, he offered the following (now empirically refuted) example to argue that Shakespeare's writings *could* have come about by pure chance:

> if you put 100,000 monkeys in a room with typewriters, given enough time one of them would produce *Hamlet*. It would have to be a very long time, of course, but the laws of chance are such that it would happen. And the universe has been around a very long time, indeed, such that if the whole history of time were thought of as one day, the "Big Bang" happening at one millisecond after midnight, *[H]omo sapiens* would not show up on this planet until

---

70   Smith, 2006, p. 3.
71   Smith, 2006, p. 3
72   Smith, 2006, p. 3.
73   Smith, 2006, p. 3

roughly 11:59 p.m.[74]

Thus, using the exact same approach that atheists use to defend their atheism, Rev. Smith concludes that no reason exists to believe in a creator God. No wonder people are leaving churches like Rev. Smith's in droves![75] Rev. Smith not only condemned the God as creator worldview, but condoned Judge Jones'[76] infamous ruling in the Kitzmiller v. Dover trail, concluding that, as a pastor, he did not

> need to spend any time defending the judge's decision to this congregation. Many of you are scientists, and educators, and thinking people who recognize foolishness when you see it … as Judge Jones did, that "Intelligent Design" is really "creationism" in a new guise, stated a bit differently, perhaps, allowing for evolutionary changes in life forms, not necessarily holding to the literal six-day creation doctrine of fundamentalists, not even specifically naming God as the Intelligent Designer. But still, a faith-based (and I would say fear-based) challenge to the theory that nearly all reputable scientists accept [namely evolution].[77] [78]

God, he explains, allowed evolution a free hand to produce whatever it may, adding that "I don't believe God is a micro-manager," but rather God is a

> gambler and risk-taker, not knowing where it will lead or how it will all turn out—there is the element of freedom, and choice, and randomness, and surprise built in. And God is even bigger than all that—more than the designer of the process, God IS the process.[79]

In other words, God is evolution, and evolution is God. One can only wonder where Rev. Smith found scriptural affirmation for this view. He provides no scientific or rational support for this idea except his weak attempt to meld atheism with his impersonal and, at best, deistic God. A leading scholar of Darwin, Harvard Professor Janet Browne, wrote that Darwin

> reinterpreted all the myriad contrivances of the living world as an

---

74 Smith, 2006, p. 3.
75 Shiflett, 2005.
76 Smith, 2006, p. 3..
77 Smith, 2006, p. 3.
78 For more on the Judge Jones decision in the Kitzmiller case, see Kevin Wirth's intro to this book.
79 Smith, 2006, p. 3.

inevitable consequence of chance and change. Now Darwin firmly drove the idea of God out of nature. As he was the first to recognize, his theory bleakly signaled the death of Adam.[80]

This reminds one of a famous debate between scientists and theologians. The scientists, Dr. Charles Signorino, Professor of Organic Chemistry, Dr. Allen Davis, Professor of Biology and Genetics, and Dr. Gary Parker, Professor of Biology, were all on the creation side of the debate. The non-scientists, Glen Koch and Robert Shinn, both religion professors and Carl Saalbach, a professor of sociology, were all on the evolutionist side.[81] It is not as rare as one might assume for the religion professors to side with evolution and the science professors to side with creation. Dembski and Richards, both ID advocates, concluded that liberal Christianity has

> great difficulty regenerating itself. Hardly anyone converts from agnosticism to liberal Christianity. Many liberal Christians started out as evangelicals. Indeed, liberal Christianity is parasitic. To survive it must recruit evangelical Christians.[82]

They add the key

> recruiting ground is the theological seminary. ...apologetics remains sufficiently unpopular at the mainline seminaries that funding, which is readily given to other campus groups, tends to get diverted from evangelical students engaged in apologetics. For instance, the *Princeton Theological Review* would long be defunct were it not for subscriptions by supporters outside the seminary as well as for donations by the students themselves. ...Standing up for Christian orthodoxy at a mainline seminary is a quick way to lose friends and alienate people. Members of the Charles Hodge Society [apologetic club] were threatened with two lawsuits for their work on the *Princeton Theological Review*, threatened with physical violence, accused of racism and sexism, denied funding that other campus groups readily received, had posted signs destroyed and removed, and were explicitly informed by faculty that membership in the Charles Hodge Society jeopardized their aca-

---

80 Browne, 1995. p. 543.
81 McKenna, 1968.
82 Dembski and Richards, 2001, p. 26.

demic advancement.[83]

One other example includes Ivan Parlor who

> had intended to become a priest, like his father, in the Russian Orthodox church. Then he discovered Darwin. It was the late 1860s, and Ivan and his brother, Dmitry, were studying at the seminary in Ryazan, where the Pavlov's lived. Early in the morning, the story goes, Ivan would sneak into the village library to read the recent Russian translation of *On the Origin of Species*.[84]

In my atheist college days we said the churches were full of useful idiots that were digging their own grave by siding with Darwinism. This was one area in which my atheist friends were correct.

12. *Many who turned to Darwinism become evolution evangelists*.

Many individuals who came from a Christian background and converted to Darwinism felt incensed at being misled by creationism and often even Christianity. As a result they devoted their lives to rescuing others from a theistic-based framework. The result in some of these cases was the formation of a life goal to show the fallacies of creation and to champion evolution. A case where this has occurred is illustrated in an article titled "Creationists Made Me Do It,"[85] which includes the following account:

> I was always a mediocre student, especially in high school. I never really knew what I wanted to do, and nothing seemed to excite me. This changed in my senior year, when a creationist visited my biology class.
>
> On that fateful day, all the science students were herded into the school auditorium, where we listened to a long and richly illustrated lecture describing literal creationism. We were informed that in an effort to "balance" our education, we would soon hear an equally long lecture on evolution. This, like many things I heard that day, turned out to be false. The evolution lecture never materialized. Remarkably, I graduated from senior biology having

---

83 Dembski and Richards, 2001., p. 26.
84 Johnson, 2008, p. 123.
85 "Creationists Made Me Do It," 2009. *Science*, 325(5943):945, August 21.

learned only about creationism.

Nonetheless, due to that lecture, he adds,

> School had finally gotten my full attention. I wanted to know what we were missing, and why. For the first time in my life, I willingly (eagerly even) picked up my textbook and studiously read it. With growing interest, I realized that evolution made an awful lot of sense, and that I was being hoodwinked by my [creation] biology class.[86]

This experience, causing a reaction which resulted in rejecting creationism, is repeated thousands of times due to poor education on this subject in the home, churches, Christian schools and even the Christian media.

An all too common example of the ramifications of Darwinism is Darrell Lambert, who sued the Boy Scouts because they require members to voice belief in God. He explained that he became an atheist as a result "of studying evolution in the ninth grade" and now wants to do what he can to persuade others to accept his view.[87]

13. *Hollywood, science fiction books, and the media promote evolution*.

Outside of the courts, academia, and the scientific community, the support for evolution permeates our western culture in scores of books, documentaries, movies, popular science magazines and news reports. These media and entertainment outlets often promote false and distorted notions about both creation and evolution viewpoints, and then fawn over the latest report of the newest scientific discovery, framed in evolutionary assumptions, without bothering to question or investigate the reliability of the stories. Most of us don't stop to think about how widespread the influence of evolution extends into our culture, nor do most of us appreciate the impact it has had.[88]

The fact is "no individual has had such a sweeping influence on

---

86 Canadian Institute for Advanced Research, Botany Department, University of British Columbia, Vancouver, BC V6T 1Z4, Canada. Patrick J. Keeling E-mail: pkeeling@interchange.ubc.ca
87 Anonymous. 2002. "Atheist Scout told to get God or Get out." *The Toledo Blade*. November 3, Section A, p. 3.
88 The endorsement by Willard Lake at the beginning of this book provides us with an idealized glimpse of how much different our world would be without the effects Darwinian evolution since *Origin of Species* was first published.

so many facets of social and intellectual life as Charles Darwin."[89]

One of the most successful Hollywood movie franchises is Michael Crichton's film adaptation of Jurassic Park, where the actors speak about evolution matter-of-factly. Other films with evolutionary assumptions and misinformation include Waterworld. In this case even evolutionists recognized the major flaws in the film where actor Kevin Costner was depicted as

> a mutant, possessing gills that allow him to breathe underwater and webbed feet that allow him to run faster… His mutation is simply an evolutionary adaptation brought about by the changing environment on Earth. In the movie the mutation may not be as beneficial as it seems as people try to kill him because of it… The concept of adaptability in evolution is one of the most misunderstood things about the theory and it's one that we see over and over again in the movies. **Organisms do not adapt based on need**, rather traits in a population are favored through natural selection. In other words, Kevin Costner wouldn't have gills because they mutated out of some necessity, the gills would have had to randomly mutate first.[90]

Another very popular series, X-men, presents the notion of powerful human mutants in various alleged emerging states of evolution. Yet one critic of this film chastises the movie using his own brand of misinformation when he writes

> The film also misuses mutations. As any evolutionary biologist or parent of a child with Down's Syndrome will tell you, mutations are almost always harmful to the organism. However, given billions of years and an unfathomable number of mutations, the vast array of species we know today have been born. **Mutations cannot cause new species to appear**, rather the appearance of the mutation and its transference to future generations will create a new species over time.[91]

The real myth lies in the critics' assumption that mutations are capable of creating "new species over time." This has never been

---

89 Padian, 2008.
90 buyDemocracy, 2012.
91 buyDemocracy, 2012.

demonstrated in the macroevolutionary sense, and most scientists today know that micro-mutations cannot be extrapolated to prove that macromutations will result from small micro adaptations, which are already encoded in the DNA of every organism. And, as noted previously, mutations are always harmful or neutral to any organism, so the real myth here is not just what was filmed in the movie, but the assumption by the critic about the notion what mutations plus time can accomplish. Unfortunately, the producers of these films erred in their depiction of some aspects of evolution, which only contributed in various ways to promote misunderstandings about how evolution works.

Hundreds of documentaries where evolution is either assumed or directly addressed can be found on any number of cable TV channels and are circulated in our public schools as resources for science students to consume. A sampling of these documentaries includes "Becoming Human, Walking With Cavemen," "The Journey of Man," "Evolution: A Journey into Where We're From and Where We're Going," and of course "Life on Earth," with David Attenborough. Taken together, these and many others in this genre form an impressive collection designed to promote the wonder of life in the context of evolution rather than in its rightful place as the wonder of the Designer's incredibly rich imagination.

The Case of Dan Brown

In many instances, someone with an evolutionary worldview can produce an enormous cultural impact. Author Dan Brown created the dark world of *The Da Vinci Code*. The book, first published in 2003, sold more than 80 million copies by 2009, and the 2006 movie starring Tom Hanks grossed over $758 million. As of 2012, his books have been translated into 56 languages and sold over 200 million copies. They also generated enormous controversy: Catholic Church leaders denounced its heretical slant and negative portrayal of *Opus Dei*, a conservative Roman Catholic group. It was obviously very anti-Catholic and featured many long-refuted claims. Brown's newest novel, *The Lost Symbol*, tells his story in his own words as follows:

> I was very religious as a kid. Then, in eighth or ninth grade, I studied astronomy, cosmology, and the origins of the universe. I remember saying to a minister, "I don't get it. I read a book that said there was an explosion known as the Big Bang, but here it

says God created heaven and Earth and the animals in seven days. Which is right?" Unfortunately, the response I got was, "Nice boys don't ask that question." A light went off, and I said, "The Bible doesn't make sense. Science makes much more sense to me." And I just gravitated away from religion.[92]

Brown did admit in the interview that "The irony is that I've really come full circle. The more science I studied, the more I saw that physics becomes metaphysics and numbers become imaginary numbers. The farther you go into science, the mushier the ground gets."

Finally, the news media typically supports any reports of alleged evolutionary discoveries without vetting them critically. In essence, the media has become the mouthpiece of evolutionary propaganda – promoting it wholesale to the general public.

* * * *

My Summary

All of the patterns mentioned here work together to orchestrate a ubiquitous reassurance that life is the product of a mindless and chance process. Like a time-tested recipe, designed to combine some of the most accessible areas of influence in western society, all of these factors comprise a formidable and pervasive cultural backdrop that is designed and reinforced to infuse the public with a unified message that evolution is a fact.

The result is that what is seen as the obvious truth of evolution is so powerful that resistance to it perceived by many to be a mark of ignorance. It's important to note that the acceptance of evolution is more a product of the cultural patterns and influences noted above, and a ubiquitous acceptance among individuals who have been indoctrinated via a one-sided education than it is due to a deep dive into the evidence and arguments for the theory. A closer examination and analysis of the flaws in the alleged evidence for evolution would result in it having far less influence and impact than we see today.

---

92 White, 2012. p. 182.

References

Andrews, Roy Chapman. 1945. *Meet Your Ancestors.* New York: Viking Press.

Anonymous. 1991. Who was Java Man? *Creation Ex Nihilo* 13 (3):22–23 June.

_____. 2010. "Stephen Hawking Says God Did Not Create the Universe: What Do You Think?" *ABC News*, September 2nd http://abcnews.go.com/print?id=11542128

_____. 2012. "Hawking at 70." *New Scientist*, pp. 26-27, January 7.

Baba, Hisao, Fachroel, Yousuke Kaifu, Glen Suwa, Reiko T. Kono and Teuku Jacob. 2003. "*Homo erectus* Calverium from the Pleistocene of Java." *Science.* 299(5611):1384-1388.

Behe, Michael. 1996. *Darwin's Black Box: The Biochemical Challenge to Evolution.* New York: Free Press

" …when Michael and I engaged in debate at the 1995 meeting of the American Scientific Affiliation, I argued that the 100% match of DNA sequences in the pseudogene region of beta-globin was proof that humans and gorillas shared a recent common ancestor. To my surprise, Behe said that he shared that view, and had no problem with the notion of common ancestry. Creationists who believe that Behe is on their side should proceed with caution - he states very clearly that evolution can produce new species, and that human beings are one of those species." From Kenneth R. Miller's review at NCSE, first published in the *Creation Evolution Journal*, Volume 16. https://ncse.com/library-resource/review-michael-behes-darwins-black-box

_____. 2007. *The Edge of Evolution: The Search for the Limits of Darwinism.* New York: Free Press.

"What is perhaps most remarkable about *The Edge of Evolution* is how much Behe now concedes to the evidence that supports Darwinian evolution. He not only accepts that life has existed on Earth for billions of years, but that it has evolved over time. He now agrees with the Darwinian notion that all life on the planet "descended with modification from one stage to another." He even acknowledges that natural selection is the obvious mechanism by which adaptive gene variants spread through a population. It is difficult to imagine his core audience being receptive to this revised position. But at this point, Behe is stuck between the need to establish a semblance of scientific credibility and the desire to forward his distinctly unscientific creationist ideas… Behe's new thesis is that there are limits to what Darwinian evolution can accomplish…Behe believes that random mutation, coupled with natural selection, is not a sufficiently powerful engine to drive the evolution of complex subcellular structures and molecular machines. Most of the really important mutations, he insists, must have been directed by an intelligent agent." From David E. Levin's review at NCSE. https://ncse.com/library-resource/review-edge-evolution

_____. *Darwin Devolves: The New Science About DNA That Challenges Evolution.* (2019) *in press.*

"…the core argument of the book …centers on the empirical data gleaned from the most thorough studies of evolution on the molecular level. Such research has only become possible in the last twenty years since new technology has enabled sequencing of DNA on large numbers of organisms. For the first time, evolutionary claims can be properly tested, and Behe presents the most rigorous analysis to date based on hard data." From the review by Brian Miller in *Evolution News,* November 21, 2018.

https://evolutionnews.org/2018/11/michael-behes-darwin-devolves-topples-foundational-claim-of-evolutionary-theory/

Bergman, Jerry. 2005. "Darwinism and the Deterioration of the Genome." *CRSQ.* September. 42(2):104-114.

_____, and Doug Sharp. 2008. *Persuaded by the Evidence* (editor). Green Forest, AR: Master Books.

_____. 2012a. *Hitler and the Nazis Darwinian Worldview: How the Nazis Eugenic Crusade for a Superior Race Caused the Greatest Holocaust in World History.* Kitchener, Ontario, Canada: Joshua Press.

_____, with Kevin With (editor and contributor) 2012b. *Slaughter of the Dissidents: The Shocking Truth About Killing the Careers of Darwin Doubters* (Volume I of the *Slaughter of the Dissidents* Trilogy). Southworth, WA: Leafcutter Press. 2nd edition.

_____. 2014. *The Darwin Effect. Its Influence on Nazism, Eugenics, Racism, Communism, Capitalism & Sexism.* Green Forest, AR: Master Books.

_____. 2015. *The Dark Side of Darwin.* Green Forest, AR: New Leaf Press.

_____. 2017. *C. S. Lewis: Anti-Darwinist: A Careful Examination of the Development of His Views on Darwinism.* Eugene, Oregon: Wipf & Stock Publishers.

_____. 2017. *How Darwinism Corrodes Morality: Darwinism, immorality, abortion and the sexual revolution*. 2017. Ontario, Canada: Joshua Press.

_____. 2019. *Science is the Doorway to Christianity.* Spokane Valley, WA: Leafcutter Press.

Berquist, Thomas H. 2012. "Peer Review: Is the Process Broken?" *American Journal of Roentgenology* (AJR), August 2012, 1992). (Editorial)

Blancke, Stefaan, Hans Henrik Hjermitslev and Peter C. Kjaergaard. 2014. *Creationism in Europe.* Baltimore, MD: Johns Hopkins University Press.

Boule, Marcellin and Henri Vallois. 1957. *Fossil Men.* New York: The Dryden Press.

Bowden, M. 1977. *Ape-Man—Fact or Fallacy: A Critical Examination of the Evidence.* Bromley, KY: Sovereign Publications.

Bowler, Peter J. 2009. "Darwin's Originality." *Science,* 323:222-226, January 9.

Brace, C. Loring and Ashley Montague. 1977. *Human Evolution.* Second Edition. N.Y.: Macmillan.

Bradley, John. 1930. *Parade of the Living.* NY: Coward-McCann.

Brady, Ronald H. 1979. "Natural Selection and the Criteria by which a theory is Judged." *Systematic Biology* vol. 28, pp. 600-21.

Brentnall, John and Russell Grigg. 1996. "Darwin's Slippery Slide into Unbelief." *Creation,* 18(1):34-37.

Browne, Janet. 1995. *Charles Darwin: Voyaging A Biography.* Princeton, New Jersey: Princeton University Press.

buyDemocracy. 2012. "5 Movies that Screw the Theory of Evolution." *By-Science* (blog) July 10

https://buyscience.wordpress.com/2012/07/10/5-movies-that-screw-the-theory-of-evolution/

Calcagno, James M. 1989. *Mechanisms of human dental reduction: A case study from post-Pleistocene Nubia.* Lawrence, KS: Dept. of Anthropology, University of Kansas.

Carrington, Richard. 1963. *A Million Years of Man: The Story of Human Development as Part of Nature.* Cleveland, OH: The World Publishing Company.

Christian Research Association. 2000. *The Australian,* February 1. p. 6.

Clinton, Susan. 1997. *Reading Between the Bones.* New York: Franklin Watts.

Cohen, I. L. 1984. *Darwin was Wrong: A Study in Probabilities.* New York: New Research Publications.

Cole, Fay-Cooper. 1925. *World's Most Famous Court Trial.* (Written testimony in Scopes Trial, pp. 234-241.) Cincinnati, OH: National Book Company.

Colling, Richard. 2004. *Random Designer: Created from Chaos to Connect with the Creator.* Bourbonnais, IL: Browning Press.

Campolo, Tony and Bart Campolo. 2017. *Why I Left, Why I Stayed: Conversations on Christianity Between an Evangelical Father and His Humanist Son.* New York: HarperOne.

Carroll, Robert. 1988. *Vertebrate Paleontology and Evolution.* New York: W.H. Freeman and Company.

http://doc.rero.ch/record/200124/files/PAL_E3902.pdf

Coyne, Jerry A. 2015. *Faith Versus Fact: Why Science and Religion are Incompatible.* New York: Viking.

Csiszar, Alex. 2016. "Peer Review: Troubled from the Start." *Nature*, Volume 532, Issue 7599, pp. 306-308. April 19, 2016.

Davidson, Phil. 1995. Pithecanthropus IV: A Human Evolutionary Ancestor or an Artificial Reconstruction? *Creation Research Society Quarterly.* 31:174-178.

Dawkins, Richard. 2013. *An Appetite for Wonder: The Making of a Scientist: A Memoir.* New York: HarperCollins.

Dembski, William. 1988. *The Design Inference.* Cambridge, England: Cambridge University Press. Summaries for chapters of this book can be read online at https://www.cambridge.org/core/books/design-inference/15A5476470 27C0CE68408E29EDA17FBC

_____ and Jay W. Richards. 2001. *Unapologetic Apologetics: Meeting the Challenges of Theological Studies*, Westmont, IL:IVP Academic.

_____ and Michael Ruse. 2004. *Debating Design, From Darwin to DNA*, Cambridge, England: Cambridge University Press.

Dennett, Daniel. 2015. "Why the Future of Religion is Bleak." *Wall Street Journal.* April 26.

Dibble, Marcia. 2012. "Anthropologist Alan Rogers' Book Aims to Convince

Skeptics that Darwin was Right." *Evolution of a Scientist*, Spring, 5 pp. http://continuum.utah.edu/2012/02/evolution-of-a-scientist/.

Dubois, Eugene. 1937. On the fossil human skulls recently discovered in Java and Pithecanthropus Erectus [*Pithecanthropus erectus*]. *Man: A Monthly Record of the Anthropological Science*. 37:1-7. January.

Evans, Rachel Held. 2014. *Faith Unraveled: How a Girl Who Knew All the Answers Learned to Ask Questions*. Grand Rapids, MI: Zondervan.

Gardner, Martin. 2013. *Undiluted Hocus-Pocus: The Autobiography of Martin Gardner*. Princeton, NJ: Princeton University Press.

Gates, R. Ruggles. 1948. *Human Ancestry*. Cambridge, MA: Harvard University Press.

Gibbons, Ann. 2003. "Oldest Members of *Homo Sapiens* Discovered in Africa." *Science*, 300:1641, June 13.

_____. 2003a. "Java Skull Offers New View of Homo Erectus." *Science*. 299(5611):1293. February 28.

_____. 2006. *The First Human*. New York: Doubleday.

Giberson, Karl. 1993. *Worlds Apart; The Unholy War Between Religion and Science* by Karl Giberson. Kansas City, MO: Bacon Hill Press.

_____. 2008. *Saving Darwin: How to be a Christian and Believe in Evolution*. New York, NY: HarperOne.

_____. *Skeptical Inquirer* [September-October, 2010, 34(3):28-42.] devoted an entire issue on Giberson

_____. *Seven Glorious Days: A Scientist Retells the Genesis Creation Story*. Brewster, MA: Paraclete Press, 2012, 190 pp.

_____. *The Wonder of the Universe: Hints of God in Our Fine-Tuned World*. Downers Grove, IL: IVP Books, 2012, 216 pp.

_____ and Donald A. Yerxa. *Species of Origins*. Rowman & Littlefield [Publishing Group], 2002, 277 pp.

_____ and Mariano Artigas. *Oracles of Science*. Oxford: Oxford University Press, 2007, 273 pp.

_____ and Francis Collins. 2011. *The Language of Science*. Downers Grove, Il. Inter-Varsity Press.

Gish, Duane. 1995. *Evolution: The Fossils Still Say No!* El Cajon, CA: Institute for Creation Research. "Java Man," pp. 280-285.

Goldschmidt, Richard. 1956. *Portraits from Memory: Recollections of a Zoologist.* Seattle, WA: University of Washington Press.

_____. 1960. *In and out of the ivory tower: The autobiography of Richard B. Goldschmidt.* Seattle, WA: University of Washington Press.

Gould, Stephen Jay. 1980. *The Panda's Thumb.* New York: W. W. Norton.

_____. 1990. "Men of the Thirty-Third Division." *Natural History*, 99(4):12-24.

_____. 1993. "Men of the Thirty-Third Division" in *Eight Little Piggies*, pp. 124-137. New York: W.W. Norton.

Gross, Rachel E. 2015. "Evolution Is Finally Winning Out Over Creationism. A majority of young people endorse the scientific explanation of how humans evolved." *Slate*, November 19..

http://www.slate.com/articles/health_and_science/science/2015/11/polls_americans_believe_in_evolution_less_in_creationism.html

Gruenberg, Benjamin. 1919. *Elementary Biology.* Boston, MA: Ginn and Company.

_____. 1924. *Elementary Biology.* Boston, MA: Ginn and Company.

Hailes, Sam. 2017. "Bart Campolo says progressive Christians turn into atheists. Maybe he's right." *Premier Christianity.* September 25.

https://www.premierchristianity.com/Blog/Bart-Campolo-says-progressive-Christians-turn-into-atheists.-Maybe-he-s-right

Ham, Ken. 2011. "What Do Leaders at Christian Colleges Believe?" *Answers*, July-Sept, p. 20.

_____ and Greg Hall. 2011. *Already Compromised: Christian Colleges Took a Test on the State of their Faith and the Final Exam is in.* Green Forest, AR: Master Books.

Hawking, Jane. 2004. *Music to Move the Stars.* New York: MacMillan.

Hawking, Steven. 1988. *A Brief History of Time: From the Big Bang to Black Holes.* New York: Bantam.

_____. 2001. *The Universe in a Nutshell.* New York: Bantam.

_____ and Leonard Moldinow. 2010. *The Grand Design*. New York: Random House.

Hendricks, Scotty. 2017. "Did Einstein Pray? What the Great Genius Thought About God." *Big Think*. http://bigthink.com/articles/did-einstein-pray-what-the-great-genius-thought-about-god

Highfield, Roger. 2003. "DNA leaders call religion to account." https://www.smh.com.au/articles/2003/03/21/1047749938110.html

Howells, William. 1947. *Mankind So Far*. Garden City, NY: Doubleday.

Horton, Richard. 2000. "Genetically Modified Food Consternation, Confusion, and Crack-up." *The Medical Journal of Australia*. 172(4).

Hrdlička, Aleš. 1916. *The Most Ancient Skeletal Remains of Man*. Second Edition. Washington, DC: Government Printing Office.

Isaacson, Walter. 2008. *Einstein: His Life and Universe*. New York, NY: Simon & Schuster, Inc..

Jammer, Max. 1999. *Einstein and Religion*. Princeton, NJ: Princeton University Press.

Jean, Frank, Ezra Harrah, and Fred Herman. 1952. *Man and His Biological World*. Boston: Ginn.

Johanson, Donald and Maitland Edey. 1981. *Lucy: The Beginnings of Humankind*. New York: Simon & Schuster, Inc.

Johnson, George. 2008. *The Ten Most Beautiful Experiments*. New York: Alfred A. Knopf.

Keith, Sir Arthur. 1925. *The Antiquity of Man*. Covent Garden, WC, London: Williams and Norgate.

Kinley, Jeff. 2015. *Gifted Mind: The Dr. Raymond Damadian Story, Inventor of the MRI*. Green Forest, AR: Master Books.

Lamoureux, Denis O. 2009. *I Love Jesus & I Accept Evolution*. Eugene, OR: Wipf & Stock.

Larsen, Kristine. 2007. *Stephen Hawking: A Biography*. Amherst, NY: Prometheus Books.

Livingstone, David N. 1987. *Darwin's Forgotten Defenders: The Encounter Between Evangelical Theology and Evolutionary Thought*. Grand Rapids, MI: William B. Eerdmans.

Lubenow, Marvin L. 2004. *Bones of Contention: A Creationist Assessment of Human Fossils.* Grand Rapids, MI: Baker Books. Revised edition.

Luskin, Casey. 2009. "Does Challenging Darwin Create Constitutional Jeopardy? A Comprehensive Survey of Case Law regarding the Teaching of Biological Origins." Hamline Law Review. Vol. 32 (1)1, Winter.

https://www.discovery.org/a/11291/

MacCurdy, George. 1924. *Human Origins: A Manuel of Prehistory.* New York: Appleton.

_____. 1935. *The Coming of Man: Pre-Man and Prehistoric Man.* New York: The University Society.

Manion, Kieran. 2015 "Peer Review Is Broken." *IMMpress Magazine,* September 27, 2015. http://www.immpressmagazine.com/peer-review-is-broken/

Martin, Douglas. 2005. H. "Bently Glass, Provocative Science Theorist, Is Dead at 98." *The New York Times* Obituaries, Thursday, January 20.

McCabe, Joseph. 1912. *The Story of Evolution.* Boston, MA: Small Maynard and Company.

McCook, 2006. "Is Peer-Review Broken?" *The Scientist,* February 1, 2006.

https://www.the-scientist.com/?articles.view/articleNo/23672/title/Is-Peer-Review-Broken-/

McKenna, Daniel J. 1968. "Evolution or Creation? How Did Man Begin? Debated at College." *Philadelphia Bulletin,* November 27.

Mehlert, A. W. 1994. *Homo erectus* 'to' Modern Man: Evolution or Human Variability? *Journal of Creation,* 8(1):105–116. April.

Mellars, Paul. 1996. *The Neanderthal Legacy: An Archeological Perspective from Western Europe.* Princeton, NJ: Princeton University Press.

Mills, David. 2003. *Atheist Universe; Why God Didn't Have a Thing to do with it.* Bloomington, IN: Xlibris.

Milner, Richard. 1990. *The Encyclopedia of Evolution.* New York: Facts on File.

Moody, Paul Amos. 1953. *Introduction to Evolution.* New York: Harper & Brothers, Publishers.

Moorcock, Michael. 2010. "Book Review: 'The Grand Design' by Stephen Hawking and Leonard Mlodinow." *Los Angeles Times*, September 5.

Naef, Adolf. 1926. Über die Urformen der Anthropomorphen und die Stammesgeschichte

des Menschenschädels, *Naturwissenschaften*. Vol. 14(21): 472-477.

Northe, Gail (Editor). 1993. *God is Evolution, Evolution is God* Cohasset, MA: Vedanta Centre Publishers.

Numbers, Ronald L. 2006. *The Creationists: From Scientific Creationism to Intelligent Design*. The Expanded Edition. Cambridge, MA: Harvard University Press.

Olasky, Marvin. 2011. "Books of the Year: Two New Books are Important Responses to the Rapidly Growing Promotion of Theistic—Or, More Properly, Deistic—Evolution." *World*, pp. 37-41, July 2.

Osborn, Henry Fairfield. 1936. *The Hall of the Age of Man*. New York: The American Museum of Natural History.

Padian, Kevin. 2008. "Darwin's Enduring Legacy." *Nature*, 451:632-634, February.

Parker, Steve. 2005. *The Dawn of Man: A Fascinating Visual Account of the Emergence and Evolution of Earth's Dominant Species*. London: Quantum Books.

Paulson, Steve. 2007. Interview of Ron Numbers: "Seeing the Light of Science." http://www.salon.com/books/int/2007/01/02/numbers/.

Perloff, James. 2001. "Time Magazine's New Ape-Man." *Creation Matters* 6(4):1-4.

Provine, William. "Response to Johnson Review." *Creation/Evolution*, Issue No. 32, Summer, 1993, pp. 62-63.

Pubpeer, the founders of. 2014. (Opinion) "Scientific Peer Review Is Broken. We're Fighting To Fix It With Anonymity." *Wired*, December 10, 2014 (Opinion).

Raymo, Chet. 1997. *Honey from Stone*. Kerry, Ireland: Brandon.

_____. 1998. *Skeptics and True Believers*. New York: Walker.

_____. 2004. *Climbing Brandon: Science and Faith on Ireland's Holy Mountain*. New York: Walker.

_____. 2006. *Walking Zero: Discovering Space and Time Along the Prime Meridian*. New York: Walker.

_____ and Maureen E. Raymo. 2001.*Written in Stone: A Geologic History of the Northeastern United States*. New York: Black Dome Press.

Regal, Brian. 2004. *Human Evolution: A Guide to the Debates*. Santa Barbara, CA: ABC Clio.

Rice, Stanley. 2007. *Encyclopedia of Evolution*. New York: Facts on File.

Rice, Stanley A. 2104. "Confessions of an Oklahoma Evolutionist: The Bad, the Ugly, and the Good." *Reports of the National Center for Science Education*, Jan-Feb.

Ridley, Mark. 1996. *Evolution*. Cambridge, MA: Blackwell.

Rissler, Leslie J. 2014. "The relative importance of religion and education on university students' views of evolution in the Deep South and state science standards across the United States." *Evolution: Education and Outreach*. 7:24, 10 September.

https://evolution-outreach.biomedcentral.com/articles/10.1186/s12052-014-0024-1

Rogers, J. Speed, Theodore H. Hubbell and C. Francis Byers. 1952. *Man and the Biological World*. New York: McGraw-Hill.

Rossiter, Wayne. 2015. *Shadow of Oz: Theistic Evolution and the Absent God*. Pickwick Publications, an imprint of Wipf and Stock Publishers.

Rusch, Wilbert H. 1991. *Origins: What is at Stake?* Creation Research Society.

Sacks, Jonathan. 2011. *The Great Partnership: God, Science and the Search for Meaning*. London, UK: Hodder & Stoughton.

Sanford, John. 2014. *Genetic Entropy*. New York: FMS Publications, 4th edition.

Schaefer, Henry F. 2003. *Science and Christianity: Conflict or Coherence?* Athens, GA: The University of Georgia.

Seeger, Raymond J. 1982. "Einstein, Cosmotheist." *Journal of the American Scientific Affiliation*, pp. 42-44, March.

Service, Robert F. 2017. Keeping the faith. *Science*. 357(6365):1088-1091,15 September.

Sheridan, Michael. 2010. "Stephen Hawking in 'The Grand Design': God Did Not Create the Universe." *Daily News*, Thursday, September 2nd.

Shermer, Michael. 1997. *Why People Believe in Weird Things*. New York: Freeman.

_____. 2000. *How We Believe: The Search for God in an Age of Science*. New York: Freeman.

_____. 2013. "Is God Dying?" *Scientific American*, December.

Shiflett, Dave. 2005. *Exodus: Why Americans are Fleeing Liberal Churches for Conservative Christianity*. New York: Sentinel.

Shipka, Tomas. 1987. "The Crisis in Peer Review," *NEA Advocate*, Apr.-May, pp. 6-7.

Shipman, Pat. 2001. *The Man Who Found the Missing Link: Eugene Dubois and his Life Long Quest to Prove Darwin Right*. New York: Simon & Schuster, Inc.

Simpson, George Gaylord. 1964. *This View of Life: The World of an Evolutionist*. New York: Harcourt, Brace and World, Inc.

_____. 1978. *Concession to the Improbable: An Unconventional Autobiography*. New Haven and London: Yale University Press.

_____. 1987. *Simple Curiosity. Letters from George Gaylord Simpson to his Family 1921-1970*. Berkeley, CA: University of California Press.

Smith, Rev. Rich. 2006. A sermon titled "100,000 Monkeys" given on January 8, 2006, at the Bethesda, MD, Westmoreland Congregational United Church of Christ.

Smith, Richard. 2006. "Peer review: a flawed process at the heart of science and journals," *Journal of the Royal Society of Medicine*. 2006 April, 99(4):178–182.

Sodera, Vij. 2003. *One Small Speck to Man: The Evolution Myth*. West Sussex, United Kingdom: Vij Sodera Productions.

Spencer, Frank. 1997. *History of Physical Anthropology*. Vol 1. New York: Garland.

Strandberg, Todd. n.d. "Evolution vs Creation: A Pointless Debate." Rapture Ready (blog)

Strickberger, Monroe. 2000. *Evolution*. Boston: Jones and Bartlett.

Stringer, Chris and Peter Andrews. 2005. *The Complete World of Human Evolution.* New York: Thames and Hudson.

Swisher, Carl C. III; Garniss H. Curtis, Roger Lewin. 2000. *Java man: How Two Geologists' Dramatic Discoveries Changed our Understanding of the Evolutionary Path to Modern Humans.* New York: Scribner.

Tattersall, Ian, Eric Delson, and John Van Couvering (editors). 1988. *Encyclopedia of Human Evolution and Prehistory.* New York: Garland Publishing.

_____, and Jeffrey Schwartz. 2000. *Extinct Humans.* New York: Westview Press.

Theunissen, B. 1989. *Eugene Dubois and the Ape-Man from Java.* Dordrecht, The Netherlands: Kluwer Academic Publishers.

Trinkaus E. and P. Shipman. 1992. *The Neandertals: Changing the Image of Mankind.* New York: Alfred E. Knopf.

Trottier, Justin. 2011. "A Little Too Grand?" *Skeptical Inquirer.* March/ April 2011. pp. 55-56.

Vaughn, Lewis. 2017. *Star Map: A Journey of Faith, Doubt, and Meaning.* Farmington, MN: Freethought Books.

Walsh, Chad. 1949. *C.S. Lewis: Apostle to the Skeptics.* New York: MacMillan.

Wells, Jonathan. 2011. *The Myth of Junk DNA.* Seattle, WA: Discovery Institute Press.

John West, "Nothing New Under the Sun," pp. 33-52, *God and Evolution: Protestants, Catholics, and Jews Explore Darwin's Challenge to Faith*, Jay W. Richards Ed. (Discovery Institute Press, 2010).

Wheaton, David. 2005. *University of Destruction Your Game Plan for Spiritual Victory on Campus.* Bethany House Publishers.

White, James. 2012. *A Traveler's Guide to the Kingdom: Journeying Through the Christian Life.* Downer's Grove: IL, Intervarsity Press.

White, Michael and John Gribbin. 1992. *Hawking; A Life in Science.* New York: Dutton.

Wiker, Benjamin. 2011. *The Catholic Church & Science; Answering the Questions, Exposing the Myths.* Charlotte, NC: Saint Benedict Press.

Wilder, T. E. 1991. "The Seventh Day: Against Humanistic Biblical Interpretation." *Contra Mundum*, 1:37-46, Fall.

Winchester, A. M.1962. *Biology and its Relation to Mankind*. Princeton, NJ: D. Van Nostrand.

Wirth, Kevin H. 2019. *Slaughter of the Dissidents Compendium*, Leafcutter Press, Chapter 1 – A Context for Discrimination Against Darwin Skeptics. (*in press*).

Weiner, Jonathan. 1994. *The Beak of the Finch: A Story of Evolution in Our Time*. New York: Knopf.

Woodruff, Lorande Loss. 1948. *Foundations of Biology*. New York: MacMillan.

Yudell, Michael. 2014. *Race Unmasked: Biology and Race in the Twentieth Century*. New York: Columbia University Press.

Zingaro, John. 2008. *Who are the Faithful? The Struggle for Truth Against Fundamentalism*. San Bernardo, CA: Published by the author.

Chapter 1

Charles Darwin:

From Creationist to Evolutionist

One of the earliest examples of a creationist becoming an evolutionist is Charles Darwin himself. As Brentnall and Grigg wrote, "Charles Darwin's thinking and writing on the subject of evolution and natural selection caused him to reject the evidence for God [that exists] in nature and ultimately to renounce the Bible, God, and the Christian faith."[93]

As a youth, Darwin wrote that "as I did not then in the least doubt the strict and literal truth of every word in the Bible, I soon persuaded myself that our [church] Creed must be fully accepted."[94] Even as a young man, Darwin wrote that when "in doubt I prayed earnestly to God to help me, and I well remember that I attributed my success to the prayers."[95] Furthermore, during his three years of theological study at Christ's College, Cambridge, where he studied to become a clergyman, Darwin

> was greatly impressed by Paley's *Evidences of Christianity* and his *Natural Theology* (which argues for the existence of God from design). He recalled, 'I could have written out the whole of the "Evidences" with perfect correctness, but not of course in the clear language of Paley,' and, 'I do not think I hardly ever admired a book more than Paley's "Natural Theology." I could almost formerly have said it by heart.' In a letter of condolence to a bereaved friend at that time, he wrote of 'so pure and holy a comfort as the Bible affords,' compared with 'how useless the sympathy of all friends must appear.'[96]

Despite his strong religious influences and his faith in God as a youth, the decline of

> Darwin's faith began when he first started to doubt the truth of the first chapters of Genesis. This unwillingness to accept the Bible as meaning what it said probably started with and certainly was

---

93 Brentnall and Grigg, 1996, p. 34.
94 Brentnall and Grigg, 1996, p. 34.
95 Brentnall and Grigg, 1996, p. 34.
96 Brentnall and Grigg, 1996, p. 34.

greatly influenced by his shipboard reading matter—the newly published first volume of Charles Lyell's *Principles of Geology* ... This was a revolutionary book ... [that] subtly ridiculed belief in recent creation in favour of an old earth, and denied that Noah's Flood was world-wide; this, of course, was also a denial of divine judgment.[97]

One of Darwin's biographers concluded that his "reading of this book was his [Darwin's] 'point of departure from [religious] orthodoxy.'"[98] And when Charles Lyell, who argued in his writings against the Genesis creation account, died in 1875, Darwin said that he "never forgot that almost everything which I have done in science I owe to the study of his [Lyell's] great works."[99]

In his *Autobiography*, Darwin wrote that he "gradually came by this time, i.e. 1836 to 1839, to see that the Old Testament was no more to be trusted than the sacred books of the Hindoos or the beliefs of any barbarian."[100] In reply to a Christian correspondent, in 1880 Darwin wrote, "I am so sorry to have to inform you that I do not believe in the Bible as a divine revelation, & therefore not in Jesus Christ as the Son of God."[101] One tragic result

of Darwin's rejection of the Bible was his loss of all comfort from it. The hopeless grief of his later letters to the bereaved contrasts sharply with the earlier letter of condolence ... In 1851, his dearly loved daughter Annie, aged 10, died from what the attending physician called a 'Bilious Fever with typhoid character.' Charles was devastated, and wrote, 'Our only consolation is that she passed a short, though joyous life.' Two years later, to a friend who had lost a child, Darwin's only appeal was to 'time,' which 'softens and deadens ... one's feelings and regrets.'[102]

The late Harvard Professor and a leading evolutionist of the last century, Ernst Mayer, wrote that before Darwin creation "was virtually unanimously endorsed by anatomists, zoologists, and botanists, including some of Darwin's closest friends, whom he later was able to convert to his views."[103] Mayer added that "It is Darwin, and Darwin alone, who deserves the credit for having

---

97 Brentnall and Grigg, 1996, p. 34.
98 Brentnall and Grigg, 1996, p. 34.
99 Brentnall and Grigg, 1996, p. 34.
100 Brentnall and Grigg, 1996, p. 277.
101 Brentnall and Grigg, 1996, p. 36.
102 Brentnall and Grigg, 1996, p. 36.
103 Mayer. 1968, p. ix.

changed this situation [from close to universal acceptance of creationism to evolution] overnight."[104] This is why evolution is often called Darwinism.

After Darwin, many hundreds of thousands of persons have followed in his footsteps, a few of which are detailed in the following chapters of this volume. I have also noted how, after leaving Christianity, many appear, in their writing at least, to be nasty and mean-spirited, especially towards their former creationist brothers. Name-calling--such as labeling creationists as pseudo-scientists, fools, intellectually challenged, retarded and worse--is very common.

References

Brentnall, John and Russell Grigg. 1996. "Darwin's Slippery Slide into Unbelief." *Creation,* 18(1):34-37.

Ernst Mayer. 1968. Introduction to a Facsimile of the first Edition of *The Origin of Species.* Cambridge, MA: Harvard University Press.

---

[104] Ibid, p. ix.

Chapter 2

Professor Warder Clyde Allee, Ph.D.

From Creationist to Atheist

Warder Clyde Allee (June 5, 1885 – March 18, 1955) is widely recognized as one of the most important pioneers of the field of ecology in America. Allee's dedication to research resulted in more than 200 papers and over a dozen books, including *Animal Aggregations: A Study in General Sociology* (1931), *Animal Life and Social Growth* (1932), *The Social Life of Animals* (1938), *Principles of Animal Ecology*, co-authored by Alfred E. Emerson, Orlando Park, Thomas Park, and Karl P. Schmidt (1949), and *Cooperation among Animals, with Human Implications* (1951).

Allee was an active biologist until he died in 1955. He was Managing Editor of the journal *Physiological Zoology*, and also chaired the *Committee on the Ecology of Animal Populations* for the National Research Council. He was also named a Fellow of the *American Academy of Arts and Sciences* in 1950, and was a Trustee for the *Marine Biological Laboratory* from 1932, until his death.

Warder's father, John Wesley Allee, was the son of a Methodist minister. At the tender age of eleven, Warder Clyde Allee claims he was officially converted at a revival meeting in Indiana. At age twenty-three, he was an undergraduate at the Quaker Earlham College. It was at Earlham where he was taught in a class by a professor who declared that "The theory and teaching that there is a God is a lie," words that hit him

> like an iron gavel hurled from a heavenless sky … There was silence in the lecture hall … Most of all he was filled with a surge of pity. He felt sorry for the misguided man, his animal evolution professor, a controversial figure who would leave his post some years later [due to] … an ugly … divorce. He had often heard of such people—infidels, atheists … but this was the first one he had met, and he planned to show him the error of his ways.[105]

In the fall of 1908, Allee had arrived as a Zoology graduate student at Hull Court, University of Chicago, then one of the few colleges that had a degree program in ecology. Warder had arrived in Chicago as a traditional Christian

---

105 Harman, 2010, p. 112.

believer, assuming that he would study the nature that God had instilled in all of His creatures. Then "Wide-eyed, he did not yet know that science would soon fix all that."[106] He researched isopods, which were "ugly little creatures" that looked "like a mysterious aquatic blend of scorpion and cricket." He was determined to crack the mystery of these tiny crustaceans that were

> abundant in shallow waters, the deep sea, and freshwater streams and ponds. Creating artificial currents in the lab, he observed that stream isopods moved toward the current more than pond ones, except when their metabolic rate was low when they were breeding. Since oxygen and carbon dioxide affected metabolic rate, and differed from streams to ponds, it had to be the gases that explained the creatures' behavior.[107]

When put into environments that contained high levels of certain metabolic

> depression agents like low oxygen, chloretone, potassium cyanide, low temperature and starvation, he could make stream isopods act like pond ones. Conversely, with high oxygen, caffeine, and elevated temperatures, lazy pond dwellers morphed into energized stream sprinters. Since all the isopods were from the same species, differences in behavior could not be due to heredity. Clearly it was all about interaction with the environment. The discovery shook his religious foundations. Hadn't the Deity instilled behavior in His creations? If so, how could coffee be so powerful? The iron gavel, he now saw, really did fall from a heavenless sky. There was no "hand of God" to behold, only physics and chemistry. Science was winning out over the supernatural.[108]

After graduating with his doctorate, Allee was increasingly disturbed by the horrific bloodshed and carnage of World War I. At the University of Chicago his interest in science could

> have laid his childhood belief in an all-powerful God to eternal rest, but at times like this, roots provided comfort. It was early 1917, and the United States still remained on the sidelines. Already conscientious objectors were being humiliated and beaten

---

106 Harman, 2010, p. 113.
107 Harman, 2010, p. 113.
108 Harman, 2010, p. 114.

in military training camps and prisons. Allee had recently been appointed Professor of Biology at Lake Forest College.[109]

He had read the Russian scientist Kropotkin's writings on the case for peace and cooperation dominating nature, and believed that Kropotkin must be correct. But he had not convinced himself of the biological justification for human cooperation with an original discovery of his own.[110] In the college chapel he preached on

> the rights of conscientious objectors. When the administration forbade him to preach in the chapel again, he spoke up in the classroom. The college docked his salary. A few faculty and students were heard murmuring the word "traitor" under their breath. Traitor? The local *Springfield News-Report* wasn't so inhibited: "Sometimes war is unavoidable, and college professors are no more necessary to civilization than carpenters and cobblers." ...If a choice had to be made, "we should prefer to give up the professors." ... At Lake Forest, Allee waited for more peaceful times. If he wanted to find scientific proof to combat the folly of human warfare, he would need to go someplace else.[111]

And he did, namely to the University of Chicago where he taught for most of the rest of his career. He may not have been an atheist, but clearly lost his earlier belief in a creator god.

Reference

Harman, Oren. 2010. *The Price of Altruism: George Price and the Search for the Origins of Kindness.* New York: Norton.

---

109 Harmon, 2010, p. 115.
110 Harmon, 2010, pp. 114-115.
111 Harmon, 2010, p. 115.

Chapter 3

Brian Alters, Ph.D.

From YEC to Aggressive Evolution Proselytizer

Brian Alters has a B.Sc. in biology and a Ph.D. in science education from the University of Southern California. He is currently Professor and Director of Chapman University's College of Educational Studies. As of this writing, he is also President of the aggressively anti-creation, anti-ID National Center for Science Education (NCSE). He has taught science education at both Harvard and McGill Universities, and is a specialist in evolution education.

With his wife, Sandra M. Alters, he has written Biology: Understanding Life which he describes as a university biology non-majors textbook. Another text, Teaching Biology in Higher Education was written to help college-level instructors persuade students of Darwinism by his very pro-evolution and anti-creation biology text. He is also the author of Teaching Biological Evolution in Higher Education: Methodological, Religious, and Non-Religious Issues, which is specifically about dealing with conflict that instructors have with students in their courses who bring up the scientific problems concerning evolution.

Alters and his wife also wrote Defending Evolution, with a foreword by late Harvard Professor Stephen Jay Gould. This book was designed to help science teachers effectively persuade readers of the truth of evolution in the light of what he admits is the evolution controversy. He also contributed a chapter to the book Not in Our Classrooms: Why Intelligent Design is Wrong for Our Schools, edited by the anti-creationist National Center for Science Education (NCSE) former President, Eugenie Scott and NCSE board member Glenn Branch.

Alters also testified as an expert witness for the evolutionists in the 2005 Kitzmiller v. Dover Area School District case as well as the retrial of Selman v. Cobb County before that case was settled out of court in favor of the plaintiff, the ACLU. Clearly, Alters has been in the forefront of fighting creationism and Intelligent Design for most of his adult life but he didn't start out that way. He grew up in a staunchly Christian environment. Writing about his experience after rejecting creationism, Alters says that when he was a creationist

"Nothing else I have done in my life has made me such an outsider," ... few of his friends or his enemies know that Alters, who had a fundamentalist Christian upbringing in southern California, rejected creationism in college. More than 2 decades later, he says, "I still have childhood friends and relatives who won't speak to me."[112]

He later explained that he became convinced of the truth of Darwinism in college, and his church, and church associations, did little to help him understand the evidence for creation and likewise, the evidence against evolution.

Alters Blames Discrimination of Creationists

In 2005, Alters was denied funding by the Social Sciences and Humanities Research Council of Canada (SSHRC) for a research project provisionally titled *Detrimental effects of popularizing anti-evolution's Intelligent Design theory on Canadian students, teachers, parents, administrators and policymakers*. One reason SSHRC listed for the rejection of his grant was the committee did not "consider that there was adequate justification for the assumption in the proposal that the theory of Evolution, and not Intelligent Design theory, was correct."

This rejection was widely reported in Nature[113] and other media.[114] [115] [116] Numerous letters were written to the SSHRC in support of Alters by the American Institute of Biological Sciences, the American Sociological Association (ASA), the Canadian Society for Ecology and Evolution, and others. The SSHRC replied by noting that the "theory of evolution is not in doubt" but claimed that the reason for the rejection was because "the committee had serious concerns about the proposed research design."

---

112 Couzin, 2008, p. 1034.
113 Hoag, 2006.
114 Secko, 2006.
http://www.the-scientist.com/?articles.view/articleNo/24111/title/No-intelligent-design--no--/
https://www.the-scientist.com/notebook-old/no-intelligent-design-no--47408
115 Prof denied grant over evolution.
http://www.canada.com/montrealgazette/news/story.html?id=0a8fac12-185a-445f-a463-5d127eaa40f2
116 Boswell. 2006. "Intelligent Design not smart enough for science," Windsor Star. April 12.

References

Boswell, Randy. 2006. "Intelligent Design not smart enough for science," *Windsor Star*. April 12.

Hoag, Hannah. 2006. "Doubts over evolution block funding by Canadian agency," *Nature*. 440(7085):720-721. April 6.

Secko, David, 2006. "No intelligent design, no $," *The Scientist*. July 1.

Video: Brian Alters - SSE *Intelligent Design on Trial* Symposium talk #5 (41:50) https://www.youtube.com/watch?v=MS0WAWt7MaU

Chapter 4

Hector Avalos, Ph.D.

From a Christian Fundamentalist to an Atheist Fundamentalist

Associate Professor of Religious Studies at Iowa State University, Hector Avalos is a former self-described fundamentalist. He is now an evangelical atheist. The former Pentecostal preacher and child evangelist is also the author of several books opposing Christianity. He has a Doctor of Philosophy in Hebrew Bible and Near Eastern Studies from Harvard University (1991), a Master of Theological Studies from Harvard Divinity School (1985), and a Bachelor of Arts in Anthropology from the University of Arizona (1982).

One major reason he left Christianity was because of evolution. He concluded from the instruction he received while in college that the creation worldview he learned about at church was false. Over time, he developed a strident opposition to anyone who is critical of evolution. A good example of his militancy was his active opposition to Intelligent Design supporter and Astrophysicist Guillermo Gonzalez, who was denied tenure at Iowa State University in 2007 because of his views. Avalos is now an internationally recognized opponent of the Intelligent Design movement. Avalos co-authored a statement against Intelligent Design in 2005 that was signed by over 130 faculty members at Iowa State University.[117] The petition stated that the claims for Intelligent Design

> are premised on (1) the arbitrary selection of features claimed to be engineered by a designer; (2) unverifiable conclusions about the wishes and desires of that designer; and (3) an abandonment by science of methodological naturalism. Whether one believes in a creator or not, views regarding a supernatural creator are, by their very nature, claims of religious faith, and so not within the scope or abilities of science. We, therefore, urge all faculty members to uphold the integrity of our university of 'science and technology,' convey to students and the general public the importance of methodological naturalism in science, and reject efforts to portray Intelligent Design as science.[118]

---

117 I could not find any information on how many professors refused to sign the statement.
118 LETTER: Intelligent Design not supported by science, *Iowa State Daily*, April 23, 3005

All of these arguments have been carefully refuted elsewhere. For example, ID does not speak to "a question of arbitrary selection of features claimed to be engineered by a designer," but the fact that clear evidence of design is everywhere in the natural world, from bacteria to humans. Furthermore, methodological naturalism is an expression used by those who attempt by definition to exclude evidence of design from being considered. We should go where the evidence leads us, without constraint by a concept such as methodological naturalism.

Avalos argued the film based on the work of Gonzalez titled The Privileged Planet should be banned because he (Avalos) incorrectly claimed that "Intelligent Design is a religious concept cloaked in the language of science"[119] when in fact Intelligent Design (ID) is a theory about known causes capable of generating complex specified information, and makes no assertions whatsoever about religion. Simply because ID may be consistent in some respects to various aspects of some religions does not make it a religious notion. Avalos added that the film is "pseudoscience," but was unable to give any evidence for this opinion, as was also true of all of the other critics that I reviewed. He rejected the film because it supports the Intelligent Design concept, which he condemned because, in his words, it supports the "old teleological argument... Design implies a Designer" whom Christians believe is God.[120]

One Iowa State University student, Scott Rank, concluded that "prominent researchers are scrambling to write articles against it [Intelligent Design], universities are firing staff members who are publicly advocating it, and Wired Magazine even devoted a cover article to it, affectionately titled 'The Crusade Against Evolution.'"[121] Rank added that "Avalos, Iowa State's most beloved atheist, argued against ID as science from the philosophical point of view, which was ironic, since Avalos is neither a scientist nor a philosopher." Rank added that "most ISU students know that Avalos will throw mud at theism whenever possible (if the ISU dietetics program hosted a Christian cooking conference, Avalos would show up with a batch of home- made Atheist cookies).["122]

Avalos' 2005 book Fighting Words: The Origins of Religious Violence used the "scarce resource theory" to explain the role of religion, specifi-

---

119 Quoted in Oltman, 2005, p. 2.
120 Hector Avalos. 2004. "The Flaws in Intelligent Design." *Iowa State Daily* (letter), October 22, p. 1.
121 Scott Rank. 2004. "Is Intelligent Design Science or Creationism." *Iowa State Daily*, October 18, p. 1.
122 Rank, 2004, p. 2.

cally Christianity, in causing violence. In short, Avalos argued that all human conflicts are usually the result of some resource that is either scarce, such as gold or silver, or perceived to be scarce, such as diamonds. He claims religion causes violence because it has created new scarce resources, such as sacred space (the "Holy Land"), group privileging, and eternal life.

Violence may result from the effort to maintain or acquire these religiously-defined resources and, Avalos argues, many people are willing to give or take life in pursuit of these resources. But we should note that fighting over scarce resources is not restricted to religious disputes; it's a behavior flaw common to all types of people, including atheists. Christianity, meanwhile, teaches the opposite: self-sacrifice and loving one's neighbor.

Unlike scarcities that are verifiable resources, eternal life is scientifically unverifiable and defined entirely by religious teachings. Christianity, his main target, offers everlasting life free to all of those who accept Christ and follow him. Thus, contrary to his claim, it is not a scarce resource at all. Unlike killing someone to obtain a resource such as money, one does not obtain heaven by taking anything away from someone else. The fact that eternal life cannot be scientifically confirmed does not establish that it does not exist. . Instead, killing someone will more likely cause the killer to lose the resource of heaven.

Avalos has even argued that religious violence is always immoral, whereas, depending on the situation, secular violence is only sometimes immoral. His book offers a scathing, but largely unsuccessful critique of the fact that the Nazi Holocaust was an example of anti-Christian inspired violence.[123]

As is true of many anti-Christian books, Avalos and his book received an enormous amount of favorable press for his very flawed thesis. On August 22, 2005, he was even featured on National Public Radio's Talk of the Nation to discuss his conversion from YEC to atheism. Similar fawning press was given to other ex-creationists including Richard Dawkins (see Chapter 12), Sam Harris, Daniel Dennett, and the late Christopher Hitchens, all of whom were former Christian creationists who became major critics of religion, especially Christianity, after adopting Darwinian evolution. His students report Avalos is a very personable and likeable teacher, which makes him very effective at turning naive students against Christianity. Some examples of his student comments include the following:

1. Dr. Avalos is very biased and intelligent. He does provide more

---

[123] See Bergman. 2012.

than one viewpoint; however, he seems to miss or mistake obvious possible explanations about the materials (Bible). Whether he does this intentionally, or on purpose it is hard to say. I feel he is a good person, but I feel bad for him because he seems to harbor a great anger towards some things that happened during his childhood church experience. Told us stories that confirmed his bitter recall of those times.

2. I took Intro to New & Old Testament classes from Dr. Avalos. He is very smart, and has most of the Bible memorized. Even on the first day he has everyone's name memorized. Class is INCREDIBLY easy. There are two tests and both are simple. He is EXTREMELY biased against all religion, but if you know that going in, he makes you think.[124]

3. He claims to be valid because he is "impartial" to subject material, yet clearly, and vocally, is against or unfavorable toward material presented. Professor Avalos has a distinct pre-supposition that orders his manner of teaching, yet refuses to recognize his own bias.[125]

4. He has all the wrong answers to all the right questions.[126]

Most of Dr. Avalos's student ratings are very good, likely because (I am told) many atheists and agnostics sign up for his classes, and he is to them their atheist/anti-Christian role model and hero.

References

Anonymous. 2007, "Intelligent Design Scientist Denied Tenure: Leading proponent of theory targeted by Atheists." World News Daily, May 12. https://www.wnd.com/2007/05/41571/
http://www.wnd.com/news/article.asp?ARTICLE_ID=55667

Bergman, Jerry. 2012. *Hitler and the Nazis Darwinian Worldview: How the Nazis Eugenic Crusade for a Superior Race Caused the Greatest Holocaust in*

---

124 http://www.ratemyprofessors.com/ShowRatings.jsp?tid=469479.
125 http://www.ratemyprofessors.com/ShowRatings.jsp?tid=469479.
126 http://www.ratemyprofessors.com/ShowRatings.jsp?tid=469479.

World History. Kitchener, Ontario, Canada: Joshua Press.

Oltman. 2005. "Film Based On Professors Book Showing at Smithsonian: Professor's Ideas Gain Recognotion in Film." *Iowa State Daily*, June 7. p.2

Video: Were the Nazis Evolutionists?: Richard Weikart vs Hector Avalos (36:38) https://www.youtube.com/watch?v=TH96XDLxkCc

Video: William Lane Craig on Hector Avalos. (10:42) https://www.youtube.com/watch?v=aM0KcIt_pXw

Chapter 5

Rev. Francisco J. Ayala, Ph.D.

From Priest to Agnostic

Rev. Francisco J. Ayala is a Spanish-American Evolutionary Biologist and Philosophy Professor at the University of California, Irvine. A former Dominican Priest, ordained in 1960, he left the priesthood at about the time that he began studying genetics. After graduating from the University of Salamanca, he moved to the U.S. in 1961 to study under the world-famous evolutionist, Theodosius Dobzhansky, for a Ph.D. at Columbia University.

Although he generally does not discuss his personal religious views, he has stated that science is compatible with religious faith in a personal, omnipotent and benevolent God, yet he once testified in court that he is not a traditional theist. His testimony in one court case is as follows[127]

Q. Do you believe that God exists?

A. It depends on what definition of God you have.

Q. What is your definition of God that exists?

A. I don't know that I am ready yet to give you a lecture in theology.

Q. I would like to know just in a summary way your conception of God.

A. Goodness in nature that can be seen as an expression of the presentation of God.

Q. There is goodness in nature. Do you think this goodness in nature has a personality?

A. No.

Q. Well, you have a Ph.D. in theology. In terms of some of the labels that are used to describe people's belief about God or the lack of a God, is there a term that

---

[127] From the Ayala deposition (Day One) in *The Association of Christian Schools International vs. Stearns, et al.* 362 Fed. Appx. 640 (9th Cir. 2010) case.

would more properly characterize your belief? Is there a term in your mind that you would characterize yourself as an atheist, an agnostic, a deist?

A. I prefer not to use any of those terms. I consider myself an independent thinker.

Q. An independent thinker?

A. Yes.

He is clearly opposed to both creationism and Intelligent Design, as is extensively documented in his many books.[128] Dr. Ayala argues that Intelligent Design is "not detectable in biological systems,"[129] and maintains that Darwin proposed "design without a designer."[130] Yet many current ID advocates challenge this assumption by pointing out that the complex information contained in the DNA of living systems is detectible scientific evidence of design.[131]

Ayala has been a Supporter of NCSE [the National Center for Science Education] since its founding,"[132] an organization founded by atheists and others to promote evolution and to combat creationists. In The Association of Christian Schools International vs. Stearns, et al. case he provided a report titled "Expert Witness Report of Francisco J. Ayala, Ph.D." Ayala wrote that "Biology for Christian Schools is not appropriate for use as the principal text in a UC-preparatory high school biology course, because students would have been taught knowledge generally rejected by the scientific community."[133] The reason this Christian textbook was rejected was because it did not teach

> evolution, which is the central organizing principle that biologists use to understand the living world. To teach biology without explaining evolution deprives students of a powerful concept that

---

[128] One reviewer of this book wrote in response to Ayala's many achievements "Tragically, however, the Scriptures make it clear that, 'the fool says in his heart, there is no God' (Psalm 14:1); and, 'the fear of the Lord is the beginning of wisdom" (Proverbs 1:7). In spite of all his degrees, awards, publications, and books, Ayala is void of true wisdom, and the Creator of all things calls him a fool."
[129] Video: Is Intelligent Design Viable? A Debate: Francisco Ayala vs. William Lane Craig (2013) at 29:50 ff
https://www.youtube.com/watch?v=mfylw5okAag&t=4s
[130] Ayala, 2007.
[131] Meyer, 2018.
[132] Branch, 2010.
[133] Expert Witness Report of Francisco J. Ayala, Ph.D In *The Association of Christian Schools International vs. Stearns, et al.* case. p. 5. Decision found here: https://web.archive.org/web/20080821201728/http://www.universityofcalifornia.edu/news/acsi-stearns/ruling0808.pdf

brings order and coherence to all biology.[134]

It is often the case that evolutionists assume Darwin skeptics who are teachers are somehow unable to accurately communicate evolutionary concepts to their students. Such assumptions serve only to further advance the false narrative that evolution critics cannot or do not adequately understand evolutionary concepts. Many teachers are lauded for their understanding of evolution until it is discovered that they advocate an intelligent designer. Then suddenly, their abilities are criticized or disputed. Yet on the other side of that coin, many evolutionists presume to accurately understand non-evolutionary views sufficiently to judge their alleged shortcomings and proclaim them inadequate or inferior to rival views.

As of this writing, Ayala now has 950 publications including 30 books. His honors take up pages, but some of the more important ones include a Fellow of the American Academy of Arts and Sciences, a member of the United States National Academy of Sciences and the American Philosophical Society. He is also a foreign member of the Russian Academy of Sciences, the Accademia Nazionale dei Lincei in Rome, the Spanish Royal Academy of Sciences, the Mexican Academy of Sciences, and the Serbian Academy of Sciences.

His honorary degrees include from the University of Athens, the University of Bologna, the University of Barcelona, the University of the Balearic Islands, the University of León, the University of Madrid, the University of Salamanca, the University of Valencia, the University of Vigo, Far Eastern National University, Masaryk University and the University of Warsaw. Such are the many rewards for his active crusade against Intelligent Design as is obvious from his many books, some of which are listed below. The best example of his support for Darwinism is his article titled: "Darwin's Greatest Discovery: Design Without Designer."[135]

Publications

His most recently published books, primarily in support of evolution and/or against ID include:

---

134 Expert Witness Report of Francisco J. Ayala, Ph.D In *The Association of Christian Schools International vs. Stearns, et al.* case. p. 5.
135 Ayala, 2007a.

Ayala, F.J. Am I a Monkey: *Six Big Questions About Evolution.* Baltimore, MD: Johns Hopkins University Press,. 2010.

_____. 2007a. Darwin's Greatest Discovery: Design Without Designer. Proceedings of the National Academy of Science, 15 May, 104:8567-8573.

_____. 2007b. *Darwin y el Diseño Inteligente. Creacionismo, Cristianismo y Evolución.* Alianza Editorial: Madrid, Spain, 231 pp.

_____. 2007. *Darwin's Gift to Science and Religion.* Washington, DC: Joseph Henry Press.

_____. 2006. *Darwin and Intelligent Design.* Minneapolis, MN: Fortress Press.

_____, and Robert Arp, eds. 2009. *Contemporary Debates in Philosophy of Biology.* London: Wiley-Blackwell. ISBN 978-1-4051-5998-2.

_____. and J.C. Avise, eds. 2007. *In the Light of Evolution: Adaptation and Complex Design.* Washington, DC: National Academy Press.

_____. and Cela Conde. 2007. Human Evolution. Trails from the Past. Oxford: Oxford University Press, 2007.

Hey, J., W.M. Fitch and F.J. Ayala, eds. *Systematics and the Origin of Species. On Ernst Mayr's 100th Anniversary.* Washington, DC: National Academies Press.

Wuketits, F.M. and F.J. Ayala, eds. *Handbook of Evolution: The Evolution of Living Systems (Including Hominids),* 2005, Volume 2. Weinheim, Germany: Wiley-VCH, 292 pp.

Ayala, F.J. Human Evolution: Biology, Culture, Ethics. In: J.B. Miller, ed., *The Epic of Evolution. Science and Religion in Dialogue,* 2004. Upper Saddle River, New Jersey: Pearson Education, Inc., pp. 166–180.

References

Ayala, F.J. 2007. *Darwin's Greatest Discovery: Design Without Designer. Proceedings of the National Academy of Science, 104:8567-8573, May 15.*

Branch, Glenn. 2010. Ayala Wins the Templeton Prize. *National Center for Science Education.* www.ncse.com/news/2010/03/ayala-wins-templeton-prize-005389

Meyer, Stephen C. 2018. Yes, Intelligent Design Is Detectable by Science. Evolution News, April 24. https://evolutionnews.org/2018/04/yes-intelligent-design-is-detectable-by-science/

Videos

Video: *Evolution of Ethical Behavior and Moral Values: Biology? Culture?* (59:36) https://www.youtube.com/watch?v=KLgKC1snRLc

Video: *Is Intelligent Design Viable?* A Debate: Francisco Ayala vs William Lane Craig   https://www.youtube.com/watch?v=mfylw5okAag

Video: *Does scientific knowledge contradict religious belief?* – 2010 (2:36) https://www.youtube.com/watch?v=AjzNWEsPr-E

Chapter 6
Edward G. Babinski
From Creationist to Agnostic

Edward Babinski became an agnostic fairly early during his intellectual journey, and an important reason was that he became convinced that Darwinism was true in college. His high school science teacher loaned him several books to read, such as Dr. Henry Morris' *The Twilight of Evolution*, which he then accepted as "Gospel truth."[136] He later decided to study biology at Christian Heritage College because of the school's close association with the Institute for Creation Research. He wanted to become a creation scientist and teach about creationism to whomever would listen in the manner of his role model, Dr. Duane Gish. Baptized as an adult believer in Christ, Babinski attended Jacksonville Chapel where, he said, he grew as a Christian. He then attended a secular college, majoring in biology. It was there where he was directly confronted with Darwinism.

At first, he continued to be active in creationism—even presenting a set of Duane Gish's slides to a college science seminar and also before a group of Ph.D. chemists at Hoffman-La Roche where he then worked in the lab. After he received his Associate Degree in biology, he transferred to Fairleigh Dickinson University in Rutherford, N.J., where he read numerous articles very critical of the Creation worldview.[137] For example, he says that, after reading "The Impossible Voyage of Noah's Ark"[138]— a "special issue of *Creation-Evolution*, a journal devoted to answering creationist claims"—"doubts crept into my fervent Creationist beliefs."[139]

Babinski then began to regularly read the anti-creationist, pro-Darwinian magazine titled *The Skeptical Inquirer*, especially those articles that would have caused him to question the validity of not only the Scriptures but also of Christianity. He concluded that, "after studying evolution's criticisms of 'scientific Creationist' arguments, I became disenchanted with Christianity in total, and became an agnostic."[140] He also has argued that the Bible and

---
136 Babinski, 1995, p. 211.
137 Babinski, 1995, p. 213.
138 Moore,,1983.
139 Babinski, 1995, p. 220.
140 Babinski, 1995, p. 221.

Christianity are based on delusions.[141] Obviously, he had a shallow knowledge of both evolution and the problems with this worldview.

Interestingly, he notes that his agnosticism has been shaken recently by the testimony of certain Christians, such as Howard Storm, and also a former hardened agnostic who is now Chairman of the Biology Department at the University where Dr. Storm taught. Babinski states that he is now no longer the skeptic he once was.[142] Nonetheless, a significant portion of his website (called Scrivenings) includes material attacking creation and attempting to prove Darwinism, such as the evidence for the evolution of whales.[143]

References

Babinski, Edward (Editor). 1995. *Leaving the Fold: Testimonies of Former Fundamentalists*. Amherst, NY: Prometheus.

Loftus, John. 2010. "The Cosmology of the Bible," a chapter in *The Christian Delusion*, Amherst, NY: Prometheus.

Moore, Robert A. 1983. "The Impossible Voyage of Noah's Ark" *Creation-Evolution* Journal Issue No. 11 4(1): 1-43.

---

[141] "The Cosmology of the Bible," a chapter in *The Christian Delusion*, edited by John Loftus (Prometheus Books, 2010).
[142] http://edward-t-babinski.blogspot.com/p/blog-page.html
[143] http://etb-whales.blogspot.com/2012/03/evolution-of-whales-adapted-from.html

Chapter 7

Robert Bakker, Ph.D.

From Creationist to Evolution Advocate

Robert T. Bakker is a world-renowned American paleontologist who helped to overthrow several modern theories about dinosaurs, arguing that dinosaurs were not stupid, clumsy, slow-moving creatures as was once almost universally thought. He was also one of the first paleontologists to suggest that dinosaurs had feathers even before any feathered dinosaur fossils had been discovered. Instead, Bakker concluded from his field research that they were loving parents that were smart, fast and adaptable. Professor Bakker was reared by creationist parents in a creationist church.[144] He grew up "in a very religious family" as a creationist and, as a child "knew the Bible well," even to the extent of convincing the draft board to give him an exemption due to his religious beliefs, which he claimed "prohibited him from taking part in any war."[145]

All of this changed when he discovered evolution at the American Museum of Natural History in New York and in college. He is now active in promoting evolution and discrediting the creation worldview, although interestingly, his talks do cover the religious background of certain famous dinosaur hunters.

Bakker graduated from Yale University where he studied under leading paleontologist professor John Ostrom, an early proponent of the new view of dinosaurs as warm-blooded animals. He then went on to Harvard to do his Ph.D. where he was exposed only to the Darwinian side in his science classes. In his career, he was a tireless researcher, and was responsible for several revolutions in paleontology, such as proving that dinosaurs were not all cold-blooded as was almost universally believed previously, but rather were warm-blooded.

As a Pentecostal preacher, Bakker is a theistic evolutionist and sees little conflict between religion and science.[146] When asked "What's the greatest enemy of science education in the U.S.? Militant Creationism?"

---

144 n.d. Interview with Bob Bakker at Prehistoric Planet. "Robert T. Bakker: Legend of Paleontology." 1999-2019. http://www.prehistoricplanet.com/features/index.php?id=26
145 Clinton, 1997, p. 87.
146 Anonymous, n.d.

> No way. It's the loud, strident, elitist anti-creationists. The likes of Richard Dawkins and his colleagues. These shrill uber-Darwinists come across as insultingly dismissive of any and all religious traditions. If you're not an atheist, then you must be illiterate or stupid and, possibly, a danger to yourself and others.[147]

Elsewhere Bakker has stated that

> The view that Religion is always the enemy of Science is simply historically false. The first universities – including Oxford – were set up by the Church. The first fossil museums – again, Oxford – were Church supported. And [the] wonderful world of Mesozoic monsters was revealed in the early 1800s by pious scholars who analyzed skeletons five days a week and then gave sermons on Sundays.
>
> The lesson of history is simple: Dinosaurs and fossils and Deep Time don't belong to Bible-bashing atheists. The wonderful pageant of prehistory is the proud legacy of evangelicals who searched for the truth in the pages of Scripture and the chapters written in stone.[148]

As a leading paleontologist, he has written numerous best-selling books espousing his views, all of which openly support Darwinism. Bakker was responsible for initiating the so-called "dinosaur renaissance" of paleontological studies, beginning with his article appropriately titled "Dinosaur Renaissance," published in the April 1975 issue of Scientific American. This article was about the new research and ideas on dinosaurs that he was involved in which proposed that they were warm-blooded. Perhaps his most popular and influential book is titled "Dinosaur Heresies (1986), which has been widely well received by both the general public and in the scientific community. He also was one of the consultants for the mega-hit film *Jurassic Park* and for the 1992 PBS series titled The Dinosaurs!

---

147 Switek, 2008.
148 Neyman, 2014.

References

Anonymous. n.d. Robert T. Bakker: Legend of Paleontology. Prehistoric Planet blog. http://www.prehistoricplanet.com/features/index.php?id=26

Clinton, Susan. 1997. *Reading Between the Bones*. New York: Franklin Watts.

Neyman, Greg. 2014. Creation Science Commentary: Dr. Bob, the Creation Scientist! Old Earth Ministries. June, 2014.

Switek, Brian. 2008. Paleontological Profiles: Robert Bakker. Laelaps Blog, April 7, 2008. http://scienceblogs.com/laelaps/2008/04/07/paleontological-profiles-rober/

Video: *Dinosaurs and the Call of Paleontology* (27:05) https://www.youtube.com/watch?v=t6NhjPANsf8

Chapter 8

Dan Barker

From Preacher to Evangelical Atheist

At the age of fifteen, Dan Barker dedicated his life to Jesus Christ. He described himself as an enthusiastic Christian, who evangelized others every chance he had. Ordained as a minister on May 25, 1975, he spent a total of seventeen years in the ministry. Then he met Darwin, and everything changed. Barker has now made a full-time career out of his change in beliefs from Christian to atheist. The reason why he left the ministry—and theism—was his five-year struggle with "reason, and intelligence, and truth" that resulted from learning Darwinian ideas.[149]

One of the first things he questioned "was the historicity of Adam and Eve," partly because, as a traveling evangelist, he learned that "some members of one congregation didn't believe Adam and Even were historical people."[150] He writes that, while still a pastor, he had a "secret life" of private reading—and was particularly impressed with science magazines. In particular, an article about "equal time for creationism" by Ben Bova in Omni Magazine

> turned the lens around so that I was gazing back at the fundamentalist mind set. The article laid bare the dishonesty of the 'equal time for Creationism in science class' argument by asking how many Christians would welcome a chapter about evolution inserted between Genesis and Exodus. I became more and more embarrassed at what I used to believe and more attracted to rational thinkers.[151]

At the far end of his theological migration, he concluded that there is "no basis for believing that a god exists, except faith, and faith was not satisfactory to me."[152] Once he came to believe that evolution was supported by science, and there existed scientific problems with the design and first-cause arguments, he reasoned that it was only logical to become an atheist. Obviously, he had a superficial knowledge of both evolution and the many prob-

---

149 Barker, 2003, p. 14.
150 Barker, 2003, p. 30.
151 Barker, 2003, p. 30.
152 Barker, 2003, p. 30.

lems with this worldview. Nonetheless, evolution turned out for him, as it has for so many others, to be the doorway to atheism.

To reject theism, Barker had, in his own mind at least, refuted all of the major arguments used to prove God's existence, focusing on the argument from design, saying that

> there is design in the universe, but to speak of design of the universe is just theistic semantics. The perceived design in nature is not necessarily intelligent. Life is the result of the mindless 'design' of natural selection. Order in the cosmos comes from the 'design' of natural regularity. There is no need for a higher explanation. The design argument is based on ignorance, not facts … God belief is just answering a mystery with a mystery, and therefore answers nothing."[153]

Barker never quite gets around to explaining the theory that the complex design everywhere in nature originated by chance mutations and natural selection, except by an appeal to the very thing he rebukes, namely ignorance and imagination.

As a music writer of some note, Barker still receives large royalties from his Christian music. He once recounted an incident with Hal Spencer, President of Manna Music, who publishes his musicals and Christian songs. Spencer was convinced of God's existence because of the design argument, and his belief that God's creation was itself strong evidence for God.[154] Now, anxious to convert his friends to atheism, Barker tried to convert Spencer by arguing that the claim "I can't imagine how that leaf got here without a Creator" can be satisfactorily explained by Darwinism—and no need exists for an intelligent creator.[155]

A key proselytizing method that he uses to convert believers into atheists is to attack Creationism. Barker states that "if a so-called 'Creation scientist' is attacking evolution, and you are not an accomplished biologist, what do you do?" Barker advises evangelical atheists to ask Creationists why they call creationism science:

> You might surprise them into the realization that they in fact have no science. And if they try to discredit evolution because it is just

---

153 Barker, 2003, p. 123.
154 Barker, 2003, p. 39.
155 Barker, 2003, p. 39.

a "theory," ask why the theory of Creation science should not be equally discredited.[156]

To prove that God does not exist, he uses the Bandwagon fallacy, an argument from authority, pointing to the fact that academicians as a group are "much less religious than the general population."[157] Although once active in evangelizing for theism, he is now very active evangelizing for atheism, including publishing, and speaking. He has been a guest on numerous talk shows, including on some of the best-known in the business, such as Oprah Winfrey. Barker is now the President of The Freedom from Religion Foundation located in Madison, Wisconsin, the largest atheist group in the U.S.[158] They have filed hundreds of lawsuits, of which he told me they have won over half. He has also been involved in over 100 debates with Christian believers. He ignores the fact that many creationists came out with a stronger commitment after graduating from a secular college. They do so well in debates that it is now difficult to find professors to debate them.

References

Barker, Dan. 2003. *Losing Faith in Faith*. Madison, WI: Freedom from Religion Foundation.

Barnhart, Melissa. "Founder of Largest Atheist Group in U.S. Shares Conservative Evangelical Background." *The Christian Post*, September 27, 2013.

---

156 Barker, 2003, p. 112.
157 Barker, 2003, p. 129.
158 Barnhart, Melissa. 2013, p. 1.

Chapter 9

Henri Bergson

An Anti-Darwinist Awarded a Nobel

Introduction

One of the leading critics of the Darwinian theory of biological origins whose work was well-received and widely accepted for years was Henri Bergson (1859-1941). He articulated some of the major problems with Darwin's theory, most of which are still problems today, especially the problem of the origin of new biological information.[159] That said, he still remained a believer in evolution throughout his life. The level of acceptance of his critique is indicated by the fact that he was nominated eight times for the Nobel and was awarded the 1927 Nobel Prize for Literature after publishing perhaps his most well-known work, *Creative Evolution*. It is the only Nobel Prize ever awarded for a critique of the Darwinian theory of biological origins.

Bergson was born in Paris, France the same year Darwin published *Origin of Species* (1859) but his family moved to London while he was still very young. As is the case with so many other young people chronicled in this volume, "in his mid-teens, he lost his faith after his acquaintance with the theory of evolution."[160] He eventually moved back to Paris as a young man, and "he matriculated into the Ecole Normale Superieure where he focused his studies on the humanities and eventually received his *licence es lettres*. He would go on to give lectures and earn his doctoral degree from the University of Paris."[161] Many of his courses focused on the theories proposed by Charles Darwin.

In the decades around 1900, several non-Darwinian theories were developed to explain the origin of new biological information, all of which now have been rejected.[162]

Bergson developed a view of Darwin's ideas around what were then regarded as the major difficulties of Darwinism. The most serious one that Bergson tried to explain was the fact that Darwinism offered no satisfactory

---

159 Margulis and Sagan, 2002.
160 Philosophical Library, http://www.philosophicallibrary.com/nobel-prize-winner-henri-louis-bergson/
161 Philosophical Library, http://www.philosophicallibrary.com/nobel-prize-winner-henri-louis-bergson/
162 Bowler, 1990.

explanation for the source of new genetic information from which natural selection could select. Bergson's theory proposed a non-Darwinian mechanism to produce new genetic information that, in turn, allowed well-documented mechanisms to function, including natural selection. [163]

Bergson's major book critiquing Darwinism was titled *Evolution Creatrice [Creative Evolution]* (1907). This 400-page bestselling book was translated into 20 languages and reprinted numerous times. The success of the book was partly because Bergson was one of the most important French philosophers of his age, and his international reputation attracted a large number of readers.[164] For his work, Bergson, Professor and Chair of the Philosophy Department at the Collége de France from 1921 to 1926, was elected to the French Academy and received the Nobel Prize in 1927. So important was his critique of Darwinism that Nobel Laureate Jacques Monod noted, when he was a youth in France "no one stood a chance of passing his baccalaureate examination unless he had read Creative Evolution."[165]

Creative Evolution Theory

The book titled *Creative Evolution* was written by Bergson to deal with what were then, and are still today, the major scientific difficulties of Darwinism. The most serious difficulty Bergson documented was the fact Darwinism offered no satisfactory explanation for the source of new genetic information from which natural selection can select.[166] Bergson was also "an unsparing critic of the [claimed] creative power of Darwinian natural selection."[167] Natural selection is not a force, but only reflects the outcome of an animal surviving, or not surviving, a change in its environment.

The key elements of Neo-Darwinism are the creation of new biological variety and the natural selection of the more fit organisms resulting from that biological variety. The survival-of-the-fittest part has been well-documented, but the *arrival* of the fittest problem, and the transformation of one species into another, were major problems in Bergson's day, and still are today. Darwin was aware his solution to this problem was merely "a provisional hypothesis or speculation," but he believed it was the best theory so far devised to explain the origin of the species. He also believed "until a better

---

163 Bothamley, 2002.
164 Fiero, 1998, p. 111.
165 Monod, 1971, p. 26.
166 Bothamley, 2002, p. 127.
167 West, 2012, p. 125.

one be advanced" his theory "will serve to bring together a multitude of facts which are at present left disconnected by any efficient cause."[168]

The "Darwinian idea of adaptation by automatic elimination of the unadapted is a simple and clear idea," but only documented Bergson's conclusion "Darwinism's reliance on accidental variation as the raw material for evolution made the development of highly coordinated and complex features found in biology nothing short of incredible. This was the case regardless of whether the accidental variations were slight or large."[169]

Bergson also documented most of the mutations which occur in life had effects that must be

> so slight that they would not hinder the survival of the organism: "For a difference which arises accidentally at one point of the visual apparatus, if it be very slight, will not hinder the functioning of the organ; and hence this first accidental variation can, in a sense, wait for complementary variations to accumulate and raise vision to a higher degree of perfection."[170]

With respect to eyesight, Bergson noted the problem with this assumption of Darwinism was, while

> the insensible variation does not hinder the functioning of the eye, neither does it help it, so long as the variations that are complementary do not occur. How, in that case, can the variation be retained by natural selection? Unwittingly one will reason as if the slight variation were a toothing stone set up by the organism and reserved for a later construction [evolution could work].[171]

Bergson noted the facts are obviously of little comfort to Darwinism, "which emphasizes that natural selection acts mechanically and without foresight. To get around this problem, other Darwinists claimed that evolution relied on large accidental variations that provided evolutionary leaps."[172] This assumption created "another problem, no less formidable," then that Bergson observed, namely

---

168 Darwin, 1896, pp. 349-350.
169 West, 2012, p. 125.
170 West, 2012, p. 126.
171 West, 2012, p. 125.
172 West, 2012, p. 125.

> how do all the parts of the visual apparatus, suddenly changed, remain so well coordinated that the eye continues to exercise its function? For the change of one part alone will make vision impossible, unless this change is absolutely infinitesimal. The parts must all change at once, each consulting the others.[173]

Bergson's observations foreshadowed Behe's irreducible complexity argument. Bergson added even "supposing chance to have granted this favor once, can we admit that it repeats the self-same favor in the course of the history of a species, so as to give rise, every time, all at once, to new complications marvelously regulated with reference to each other, and so related to former complications as to go further on in the same direction?"[174]

Among many other problems with Darwinism that Bergson attempted to address include the fact that multicellular organs are a "functional whole made up of coordinated parts," and if "just one or a few of the parts happened to vary ... the functioning of the whole would be impaired."[175] This concept, now known as Irreducible Complexity, is the basis of the modern academic movement known as Intelligent Design. Bergson also concluded, due to Irreducible Complexity, at every stage of an animal's history and development "all the parts of an animal and of its complex organs" must

> have varied contemporaneously so that effective functioning was preserved. But it is utterly implausible to suppose, as Darwin did, that such coadapted variations could have been random.... Some agency other than natural selection must have been at work to maintain continuity of functioning through successive alterations of form.[176]

In short, the "sheer improbability of the Darwinian explanation increases exponentially once one realizes how frequently the same complex biological features are supposed to have arisen independently in different evolutionary lineages."[177] In Bergson's words: "What likelihood is there that, by two entirely different series of accidents being added together, two entirely different evolutions will arrive at similar results?"[178] This idea, Bergson concluded, was

---

173 Bergson, 1944, pp. 72-73.
174 Bergson, 1944, pp. 72-73.
175 Goudge, 1967, p. 292.
176 Goudge, 1967, p. 292.
177 West, 2012, p. 126.
178 Bergson, 1944, p. 74.

incredible because an

> accidental variation, however minute, implies the working of a great number of small physical and chemical causes. An accumulation of accidental variations, such as would be necessary to produce a complex structure, requires therefore the concurrence of an almost infinite number of infinitesimal causes. Why should these causes, entirely accidental, recur the same, and in the same order, at different points of space and time?[179]

Bergson's answer to his own question was, no one should accept this theory

> and the Darwinian himself will probably merely maintain that identical effects may arise from different causes, that more than one road leads to the same spot.... [L]et us not be fooled by a metaphor. The place reached does not give the form of the road that leads there; while an organic structure is just the accumulation of those small differences which evolution has had to go through in order to achieve it."[180]

Consequently, the "struggle for life and natural selection can be of no use to us in solving this part of the problem, for we are not concerned here with what has perished, we have to do only with what has survived."[181] West writes according to

> the extensive annotations Lewis made in his personal copy of L'Evolution Creatice, it is clear that he [C. S. Lewis] understood and appreciated Bergson's critique of natural selection. Lewis aptly summarized the Darwinian mechanism of adaptation according to Bergson as the "[e]limination of the unfit" and noted that it "plainly cannot account for complicated similarities on divergent lines of evolution." Lewis also noted Bergson's view that "pure Darwinism has to lean on a marvelous series of accidents" and ... Darwinists try to "escape" this truth "by a bad metaphor." Lewis paid particular attention to Bergson's critique of Darwinian accounts of eye evolution in mollusks and vertebrates, concluding that "[n]atural selection... fails to explain these eyes."[182]

---

179 Bergson, 1944, p. 74.
180 Bergson, 1944, p. 74.
181 Bergson, 1944, p. 74.
182 West, 2012, p. 127.

Bergson used "detailed scientific arguments as well as philosophical ones" to support his view, and gained many followers among well-educated intellectuals, including Alfred North Whitehead and philosopher George Santayana.[183] For several reasons, Bergson's work did not gain many followers among biologists.

Although deeply influenced by Herbert Spencer, John Stewart Mill and Charles Darwin, Bergson's theory largely was a reaction to their philosophies and worldview. Although Bergson proposed another solution to the very real and serious problems that he identified in Darwinism, this does not detract from his insight into the major problems with Darwinism. Bergson also concluded Darwinism had failed to explain

> why living things have evolved in the direction of greater and greater complexity. The earliest living things were simple in character and well adapted to their environments. Why did the evolutionary process not stop at this stage? Why did life continue to complicate itself "more and more dangerously?"[184]

Bergson further argued the mechanism of natural selection, which only eliminates the less fit, could not answer the critical problems he (Bergson) identified because something "must have driven life on to higher and higher levels of organization, despite the risks involved."[185] That something, Bergson proposed, was the élan vital meaning the mysterious force directing evolution.

A major problem Bergson unsuccessfully tried to deal with was explaining the origin of his vital force, and demonstrating exactly how it functions to produce new life forms. Among the many other problems with Bergson's theory was the fact that it did not explain why evolution took the smooth tree-like route supporters claimed, instead of some other, more direct, path.[186]

Bergson Awarded the Nobel

Although in the public mind, Bergson's Nobel is linked directly to "his masterwork *Creative Evolution*," the Nobel award was actually for the full scope of his work, including his work in biology, and his writings in aesthetics

---

183 Bothamley, 2002, p. 127.
184 Goudge, 1967, p. 292.
185 Goudge, 1967, p. 292.
186 Bothamley, 2002, p. 127.

dealing with the implications of Darwinism.[187] The Nobel citation reads "in recognition of his [Bergson's] rich vitalizing ideas and the brilliant skill with which they are presented." His major works were published posthumously in seven volumes between 1945 and 1946.[188]

Bergson's Influence on Others

C. S. Lewis learned a lot about the many major problems of Darwinism from Bergson. One of Lewis' most heavily annotated books was Bergson's *Creative Evolution* critiquing the creative ability of Darwinian natural selection theory. Lewis first read Bergson's book "as a 19-year-old soldier during World War I while recovering from shrapnel wounds."[189] Lewis wrote many notes in the book and underlined a large number of pages, later stating the book's "critique of orthodox Darwinism is not easy to answer" and it influenced his energetic stand against materialistic Darwinism.[190]

A few years later, on Aug 14, 1925, Lewis wrote a letter to his father saying that the evolutionary ideas of both Charles Darwin and Herbert Spencer were built "on a foundation of sand." Lewis was still an atheist when he expressed these early doubts about Darwin.

Bergson's philosophical body of work, including creative evolution, was honored with a Nobel because it appeared to a large number of scholars to be a plausible explanation for the source of genetic variety and many other problems of Darwinism. When his theory of creative evolution was discredited, those who put their faith in it were forced to postulate some other mechanisms, as discussed below.

Attempts to Replace his Theory

One mechanism for evolution popular for several decades was **orthogenesis**. This theory taught evolution occurs due to the influence of internal organismic forces guiding variation into specified directions. The theory postulated organisms are driven to perfection, just as an embryo is driven to develop into an adult by internal forces. This theory also was eventually discredited when no mechanism was found that could supply the forces or direction.

Another major theory of the source of variations is macro-mutations.

---

187 Schlessinger and Schlessinger, 1986, p. 56.
188 Bergson, 1945-46.
189 West, 2012, p. 125.
190 Lewis, 1996a, p. 89.

Hugo De Vries (1848-1935) claimed he demonstrated from his research on the evening primrose that dramatic new varieties and traits could arise suddenly without explanation. He and others believed these macro-mutations finally gave evolutionists a mechanism for producing new genetic traits in plants and animals. Further research found De Vries results were not due to mutations, but to the fact that the evening primrose has unequal chromosome numbers causing the hybrid plants to *appear* to produce new varieties.

The idea of macro-mutations was briefly resurrected in the 1940s by University of California, Berkeley, geneticist Richard Goldschmidt. He concluded the origin of major new animal and plant varieties was due to single mutations involving large and complex changes that happened to be successful. Goldschmidt called such creatures "hopeful monsters."

We now know far more than one mutation would be required to produce the changes necessary to evolve a new animal order—many hundreds or thousands would usually be needed. Since these alternative theories have now been discredited, no satisfactory mechanism to produce evolution by macro-mutations has been proposed by modern Neo-Darwinists.

Today, many evolutionists assume a large number of small mutations, and not the macro-mutations as De Vries and Goldschmidt argued, can account for macro-evolution. This conclusion is not based on experimental evidence, though, but on the assumption that the evidence for micro-evolution can be extrapolated to macro-evolution. The empirical evidence, however, is clear—neither macro-mutations nor micro-mutations can provide a significant source of new genetic information as even some Darwinians admit: "Mutation accumulation does not lead to new species or even to new organs or tissues."[191]

What mutations eventually lead to is not progressive evolution, but rather sickness and death because the vast majority of mutations, over 99.99 percent, are near neutral or harmful. Margulis, then President of Sigma Xi, the honor society for scientists, added that "many biologists claim they know for sure that random mutation (purposeless chance) is the source of inherited

---

[191] Margulis and Sagan, 2002, p. 11. NOTE: Margulis and Sagan state in this article "But many biologists claim they know for sure that random mutation (purposeless chance) is the source of inherited variation that generates new species of life and that life evolved in a single-common-trunk, dichotomously branching-phylogenetic-tree pattern! 'No!' I say. Then how did one species evolve into another? This profound research question is assiduously undermined by the hegemony [of those] who flaunt their 'correct' solution. Especially dogmatic are those molecular modelers of the 'tree of life' who, ignorant of alternative topologies (such as webs), don't study ancestors. Victims of a Whiteheadian 'fallacy of misplaced concreteness,' they correlate computer code with names given by 'authorities' to organisms they never see! Our zealous research, ever faithful to the god who dwells in the details, openly challenges such dogmatic certainty."

variation that generates new species of life… 'No!' I say."[192]

Both creationists and Intelligent Design advocates conclude that the only plausible source of new life forms, as Bergson postulated, is intelligence. In each new generation of humans an estimated 100 to 200 new mutations are added to their gene line.[193] Intelligent Design only postulates an intelligent source of new mutations, but other creationists proclaim this source is an Intelligent Creator they refer to as God.

Summary

Almost a century after Bergson, Neo-Darwinists still hotly debate the source of new genetic information they believe propels macro-evolution.[194] This state of affairs has not been due to a lack of hypotheses. Theories such as "Creative Evolution," orthogenesis, macro-mutations, and others that have been "substituted for the Darwinian mechanism" have received wide support but, when carefully examined, all eventually were abandoned as untenable.[195]

So far, no post-Darwinian theory has been able to deal with the major inadequacy of Neo-Darwinism Bergson documented, namely, the source of new biological information. Indeed, the lack of compelling evidence demonstrating the origin of new species information turns out to be the Achilles Heel of evolution. Clearly, as stated by one Harvard biochemist, "evolutionary theory is a tumultuous field where many differing views are now competing for dominance."[196]

References

Bergman, Jerry. 2003. "The Century-and-a-half Failure in the Quest for the Source of New Genetic Information." *T.J. Technical Journal* 17(2):19-25.

Bergson, Henri L. 1911. *Creative Evolution.* New York: Henry Holt.

_____. 1944. *Creative Evolution.* New York: The Modern Library. Translated by Harvard University Professor Arthur Mitchell.

_____. 1945-46. *Oeuvres Completes.* Geneva, Switzerland: Albert Skira.

---

192 Margulis, 2006, p. 194.
193 Meisenberg and Simmons, 2008, p. 153.
194 Morris, 2001; Sterelny, 2001.
195 Bothamley, 2002, p. 127; Bergman, 2003.
196 Esensten, 2003, p. 2.

Bothamley, Jennifer. 2002. *Dictionary of Theories*. Canton, MI: Visible Ink Press.

Bowler Peter. 1990. *Charles Darwin: The Man and His Influence*. U.K.: Blackwell.

Darwin, Charles. 1896. *The Variation of Animals and Plants Under Domestication. Vol. II*. New York: D. Appleton.

Edman, Irwin. 1944. Foreword to Bergson's 1944 *Creative Evolution*.

Esensten, Jonathan H. 2003. "Death to Intelligent Design." *The Harvard Crimson Online Edition*, March 31, 2003.

Fiero, Gloria K. 1998. *The Humanistic Tradition. Book 5: Romanticism, Realism, and the Nineteenth-Century World*. Third Edition. New York: McGraw-Hill.

Goudge, T.A. Editor 1967. "Henri Bergson" pp. 287-295 in Volume I *The Encyclopedia of Philosophy*. New York: Macmillan.

Margulis, Lynn and Dorion Sagan. 2002. *Acquiring Genomes. A Theory of the Origins of Species*. New York: Basic Books.

Margulis, Lynn. 2006. "The Phylogenetic Tree Topples," *American Scientist*. 94(3):194.

Meisenberg, Gerhard and William Simmons. 2008. *Principles of Medical Biochemistry*. New York: Mosby Elsevier.

Monod, Jacques. 1971. *Chance & Necessity*. New York: Knopf.

Morris, Richard. 2001. *The Evolutionist: The Struggle for Darwin's Soul*. San Francisco, CA: W. H. Freeman.

Schlessinger, Bernard and June Schlessinger. 1986. *The Who's Who of Nobel Prize Winners*. Phoenix, AZ: Oryx Press.

Sterelny, Kim. 2001. *Dawkins vs. Gould: Survival of the Fittest*. (Oxford, Cambridge) London: Icon Books.

West, John G. (editor). 2012. *The Magician's Twin: C. S. Lewis on Science, Scientism, and Society*. Seattle, WA: Discovery Institute Press.

Chapter 10

Ross Blythe

Journey from Christian to Atheist

Dr. L. Ross Blythe, now Professor Emeritus from Purdue University, tells of his journey from being a Christian to becoming an atheist in the January 2019 issue of the atheist magazine *Free Inquiry*.[197] As Blythe documents, the major issue in his case was creation. His story is a good summary of almost all of the cases I have documented in this book. Blythe writes:

> Sadly, there are religious people still claiming that evolution is false and that "intelligent design" described in the Scriptures is the final authority. Our schools today are being influenced by these "creationists" who are demanding that creationism be included in the curriculum. How sad that our country is still wrestling with this problem, which is detrimental to the objectives of science.[198]

Blythe adds that "voucher" school plans are perpetuating this folly. "…It is amazing to me that even today there are people who are climate-change deniers, people who believe Earth is flat, or those who believe Earth is only about 6,000 years old."[199] Blythe's education includes four years at a Christian institution, Wheaton College in Wheaton Illinois (1952-1956). He obviously did not get a very good background in apologetics at Wheaton. He writes:

---

197 Blythe, 2018-2019.
198 Blythe, 2019. p. 55. Even more sadly, these claims represent serious misunderstandings commonly proclaimed about the intent and influence of many creationists and ID advocates. His comment reveals a poor understanding of the position taken by ID advocates. First, even though many ID notions may be consistent with biblical scripture, no religious texts are used by mainline ID advocates to promote ID arguments because none are needed. Second, the champions of ID point to specified and irreducible complexity of life arguments and the origin of information as strong indicators that living systems show clear evidence of design. Third, ID is a scientific theory, not a religious one. Finally, most ID advocates do not "demand" that their views be included in public school curriculums. In fact, they have clearly stated that this is not one of their objectives. See The Discovery Institute's FAQ page for more information, especially the section titled Questions about Science Education Policy. https://www.discovery.org/id/faqs/ Biblical creationists see the evidence of science as confirmation of what the bible teaches. This doesn't mean their view of science is "religious." Every scientist, of whatever religious or non-religious persuasion they may hold, is still able to utilize a scientific approach to evaluating evidence. If they all abide by similar rules of investigation and analysis but arrive at different conclusions, this doesn't necessarily mean they did not execute a genuinely scientific inquiry.
199 Blythe, 2019. p. 55.

> My education at Wheaton included a strong liberal arts curriculum, tainted as it was by evangelical dogma. In reality, if Wheaton were to *really* educate its students, it would be out of business, because all graduates should rationally deny the "truth" of the Scriptures and theology, in favor of all that science teaches us.[200]

Use of the word "tainted" indicates he obtained almost no grounding in apologetics. Blythe concludes: "My life has taken such a positive turn, and I am very content with my life without religion. My only regret is that my "epiphany" didn't occur many years earlier." Another reason for leaving Christianity was his experience in Korea. He writes after he graduated from Nike School at Fort Bliss, Texas, his first assignment turned out to be

> a sixteen-month tour in Korea from January 1957 to June 1958. Religion still had its grip on me, yet I was being exposed daily to the lives of Koreans suffering from desperate poverty and deprivation. This is when another revelation became part of my experience. Remembering the chorus, "Jesus loves the little children, all the children of the world …," I could not reconcile believing God's love and witnessing the suffering of the Koreans, especially their children. How could a loving god who is omniscient and omnipotent allow this?[201]

This was clearly one of Blythe's turning points, and because he was unable to find answers to the horrible conditions he witnessed, he blamed God, as many others do.[202] Blythe adds the only reason he remained a Christian at that time was because of the influence of his peers: "All of our friends were practicing Christians, so I gave little thought to the question of the existence of God." The first step in his change was, when "surfing" TV channels, he landed on

> C-SPAN, where the program was about secular humanism and the speaker was Sam Harris. My whole attention was immediately centered on Dr. Harris's comments. I immediately bought his book *The End of Faith* and read it almost in one sitting. This

---

200 Blythe, 2019, p. 55, italics in original.
201 Blythe, 2019, p. 54.
202 Sadly, Blythe lost his perspective on the clear teaching of the Bible that speaks to why the world is in its current state (due to the fall) and the plan of God to right the wrongs of evil.
https://www.bethinking.org/would-a-good-god-allow-suffering/q-why-does-god-allow-evil-to-exist

was my epiphany. I continued with his *Letter to a Christian Nation* and moved on to Richard Dawkins's *The God Delusion*, followed by Christopher Hitchens' works. Never in my life did anything so crystallize my thinking about the nonexistence of God as these books did.[203]

These books have been carefully refuted even by those who disagree with the Darwinian worldview.[204] In another case in the same issue of *Free Inquiry*, the author, geologist Alexander Schriener, writes

> I flirted with religions in my youth, it was my grounding in geology and other physical sciences that solidified me in rejecting the divine aspects of religions. The unrefutable[205] evidence of the age of the planet and the universe, evolution of life (including humans), and natural causes of geologic and physical forces that affect the planet make it clear that a divine source for such things is simply mythology.[206]

Yet another writer in the same issue, Professor R. C. Gibson adds, again stressing "reason," quotes his geologist colleague, Alexander Schriener, who claimed "There are positive philosophical principles, social support structures, and community outreach aspects of many religions that can benefit one's life and help others." Gibson answered as follows:

> "Positive philosophical principles"—really? All religions fear reason and venerate faith; they currency in closed minds, anathema to any philosophical principles … how religious people view nonbelievers—as heretics, blasphemers, and they have a special place reserved for us after death … Science (reason) is at war with religion (superstition) for the minds of human beings; progress has been slow, and the last thing I worry about is offending someone who would burn me at the stake or wish me to their version of Hell.[207]

---

203 Blythe, 2019, p. 54.
204 John Blanchard. 2010. Dealing with Dawkins. London: Evangelical Press; Ken Ammi 2017. *Pop-Atheist Bible Expositors: Featuring Richard Dawkins, Christopher Hitchens, Sam Harris, Dan Barker, and Neil deGrasse Tyson*. New York: CreateSpace Independent Publishing Platform. See also: Ward, Keith. 2011. *Is Religion Dangerous?* Oxford, England: Lion.
205 If something is "unrefutable," then it doesn't qualify as scientific since any scientific notion must always be subject to falsification and refutation as more information becomes available.
206 Schriener, 2019, p. 66.
207 Gibson, 2019, p. 66.

48 • Evolution is the Doorway to Atheism

The problem for Blythe now became exacerbated due to his existing relationships with his religious friends and family. He writes

> I am in a quandary about expressing my new-found freedom from religion and speaking out for secular humanism. It is a difficult balance, and I try to subtly make my case without alienating any of my lifelong friends ... Charles Darwin was in the same predicament that I am. If his conclusions about evolution were published, Darwin would have alienated his family—especially his wife, who was very devout. In fact, Darwin might have been tried for heresy had he made his work public immediately. He waited many years before announcing his theory.[208]

Darwin would hardly been tried for heresy—both his grandfather and father held the same beliefs, as did many others, and none of them were tried for heresy. Actually, the opposite occurred. They were lionized and rewarded not only with record book sales but honors from the scientific community.

This case supports the statement by Darwinist Professor of Philosophy Michael Ruse who correctly noted there are two kinds of evolution, the fact of evolution and the other side which unfortunately tends to dominate today, at least in many public discussions. Professor Ruse also opined that:

> I am an ardent evolutionist and an ex-Christian, but I must admit that in this one complaint ...— the literalists are absolutely right. Evolution is a religion. This was true of evolution in the beginning, and it is true of evolution still today.[209]

Some background on Ruse. He testified against creation in the court case challenging the *"balanced treatment"* bill dismissed in Louisiana back in 1987.[210] Dr. Duane Gish, who earned a PhD in biochemistry from the University of California, Berkeley commented as follows:

> "Dr Ruse," Mr. Gish said, "the trouble with you evolutionists is that you just don't play fair. You want to stop religious people from teaching *our* views in schools. But you evolutionists are just as re-

---

208 Blythe, 2019, pp. 54-55.
209 Ruse, 2000. p. B3, Wein. 2000.
210 Edwards v. Aguillard. 482 U.S. 578 (1987).
Even as late as 2016, the Louisiana senate failed on a vote to remove this law from the books.
https://www.nola.com/politics/index.ssf/2016/03/louisiana_senators_refuse_to_r.html

ligious in your way. Christianity tells us where we came from, where we're going, and what we should do on the way. I defy you to show any difference with evolution. It tells you where you came from, where you are going, and what you should do on the way. You evolutionists have your God, and his name is Charles Darwin."[211]

At the time, Ruse dismissed Gish's comments, but he later found himself

thinking about his words on the flight back home. And I have been thinking about them ever since. Indeed, they have guided much of my research for the past twenty years. Heretical though it may be to say this -- and many of my scientist friends would be only too happy to chain me to the stake and to light the faggots piled around -- I now think the Creationists like Mr. Gish are absolutely right in their complaint.[212]

This case is a good example of the failure of the American educational system to help students evaluate issues such as creationism versus evolution. Professor Blyth seems to have been a victim of the false notion that biblical references are needed to support design oriented explanations as challenges to Darwinism. If his article in the atheist magazine is any indicator, he very likely perpetuated this problem when teaching his own students. Schools do not educate, but as this book documents, indoctrinate students in a Darwinian-only worldview.

References

Ken Ammi 2017. *Pop-Atheist Bible Expositors: Featuring Richard Dawkins, Christopher Hitchens, Sam Harris, Dan Barker, and Neil deGrasse Tyson*. New York: CreateSpace Independent Publishing Platform.

John Blanchard. 2010. *Dealing with Dawkins*. London: Evangelical Press.

Blythe, Ross. 2019. "My Journey to Unbelief." *Free Inquiry*, pp. 54-55. December 2018 - January 2019.

Gibson, R. C. 2019. Religious Scientists. *Free Inquiry*, p. 66. December 2018 - January 2019.

---

211 Ruse, 2000. p. B3. Wein. 2000.
212 Ruse, 2000. p. B2. Wein. 2000.

Ruse, Michael. 2000. "Saving Darwin from the Darwinians." *National Post*, Saturday, May 13

p. B3 https://www.huffingtonpost.com/michael-ruse/is-darwinism-a-religion-b-904828.html.

Schriener, Alexander. 2019. Religious Scientists. *Free Inquiry*, p. 66. December 2018 - January 2019.

Ward, Keith. 2011. *Is Religion Dangerous?* Oxford, England: Lion

Wein, Richard. 2000. "How evolution became a religion." ASA blog post. http://www2.asa3.org/archive/evolution/200005/0160.html

Chapter 11

Bart Campolo

The Consequences of Ignoring Apologetics

One typical example of the consequences of ignoring apologetics is the case of Bart Campolo, the son of the well-known Christian college professor, Tony Campolo, who left the Church and became a secular humanist,[213] and no longer believes in God.[214] As is often true, exposure to Darwinism in school and an inability to defend his beliefs was a critical reason why Bart rejected theism. Bart writes "thanks to Charles Darwin, we have at least a pretty good idea of how it [life] moves forward from simplicity of a single celled something to the complexity of us."[215] In the end, Bart concluded "the totality of matter and energy … is all that is real."[216] He also no longer believes that there is "any overarching purpose or design" of life and "the meaning of life … is that there isn't one. In short, the universe doesn't care" about us.[217]

One can understand why Bart's father was so devastated when his son became an atheist and a nihilist. Bart reasons that, although life is a miracle, given trillions of years, stars, and planets "such a miracle was bound to happen somewhere" sooner or later, and here we are as a result.[218] Thus, God is not needed because evolution explains everything. Given this reasoning, Bart concluded, why should we believe in God? This case, and millions of others, illustrates the fact that Darwinism is the Doorway to Atheism. After explaining how mutations are the source of genetic variety that evolution selects, Bart opines that the most important DNA mutation is

> the one that separated the first of us mammals from our reptilian ancestors. After all, while reptilian brains work just fine when it comes to managing hunger… and the other basics of survival, they have no capacity for memory or emotion. We mammals, on the other hand, have added to those reptilian instincts what biologists call the limbic system, which enables us to feel emotions, re-

---

213 Campolo and Campolo, 2017, p. 139
214 Hailes, 2017.
215 Campolo and Campolo, 2017. p. 145.
216 Campolo and Campolo, 2017 p. 144.
217 Campolo and Campolo, 2017, pp. 144-145.
218 Campolo and Campolo, 2017, p. 146.

member experiences and cooperate with one another as a survival strategy.[219]

Bart adds that what sets human beings apart from animals is the development of "the prefrontal cortex, where we reason, think logically, recognize the passage of time, generalize our experiences, and make complex decisions. Our prefrontal cortexes are what enable us to extend primitive moral intuitions into universal standards of behavior—like the Golden Rule."[220] Bart doesn't think his experience is unique, but is in fact a growing trend among progressive Christians and believes "…the current world of Christianity is heading towards full blown unbelief," adding that he predicts

> as many as 40% of progressive Christians will become atheists over the next decade. In his view, the process of abandoning Christian doctrines is almost addictive. Once you start, you don't know where to stop. It might begin with "dialing down" your view of God's sovereignty, but it could easily end with unbelief.[221]

Bart also had a professor of the Old and New Testament, Dr. Barr, who spent most of the term documenting what he claimed are its many errors, contradictions, and morally repugnant passages. After his exposure to this view, Bart said "the results were devastating" to his faith.[222] Most of these claims, as is true of the other arguments Bart noted, are easy to resolve if he had done more research into apologetics. And most of the arguments that Bart listed, his father, Tony, ignored. Tony's major argument seemed to be that he really loves Jesus, and Bart should also, and that should settle it. Tony's response to evolution was, when considering evolution, we should remember

> that many scientists do not believe that natural selection is simply a process of random trial and error. … the evolutionary development of living organisms is being guided. I am not a young earth creationist, but obviously I believe that the guiding force is … God. So then, you can count me in with those oft-ridiculed religionists who claim there is an "intelligent designer" driving the creative processes of the universe.[223]

---

219 Campolo and Campolo, 2017, p. 110.
220 Campolo and Campolo, 2017, p. 111.
221 Hailes, 2017.
222 Campolo and Campolo, 2017, p. 45.
223 Campolo and Campolo, 2017, p. 139.

Tony added that he wrestles with why secularists like his son Bart choose to "live as though there is nothing and no one behind the awesome wonder of the universe or our most transcendent experiences, and then cherry-pick the theories and arguments that best support that lifestyle, [which] is an act of faith as well."[224] Bart chose this worldview as a result of his secular education and his lack of education in the opposing view.

Bart also relates that his hero, Ms. Ursula, the daughter of a minister, wrote that the scientific account of creation is the epic of evolution, which is "The Big Bang, the formation of stars and planets, the origin and evolution of life on this planet, the advent of human consciousness and the resultant evolution of cultures—this is the story, the one story, that has the potential to unite us, because it happens to be true."[225]

In short, Tony Campolo could not defend his faith, and a major reason was because he accepted evolution as God's means of creation, which seriously handicapped his ability to effectively defend his worldview. His son, Bart saw that once one accepts evolution, it is very difficult to defend the main reason most people give for believing in God, namely the argument from design.[226]

References

Campolo, Tony and Bart Campolo. 2017. Why I Left, Why I Stayed: Conversations on Christianity Between an Evangelical Father and His Humanist Son. New York: HarperOne.

Hailes, Sam. 2017. "Bart Campolo says progressive Christians turn into atheists. Maybe he's right." Premier Christianity. September 25. https://www.premierchristianity.com/Blog/Bart-Campolo-says-progressive-Christians-turn-into-atheists.-Maybe-he-s-right

---

224 Campolo and Campolo, 2017, p. 139.
225 Campolo and Campolo, 2017, p. 70.
226 Campolo and Campolo, 2017.

the marvels of the human immune system—can't be the work of random chance and natural selection. Intelligent design advocates look at these sophisticated components of living things, [and] can't imagine how evolution could have produced them and conclude that only God could have. That makes Colling see red.[230]

He concluded that "overwhelming evidence" exists that God is "truly a Random Designer."[231] His writings indicate that Colling might be closer to an agnostic, such as when he wrote "in spite of our inability to see, touch, or understand God in the physical dimension, we intuitively sense He is real—or at the very least, that He might be real."[232]

When all life was evolving by elimination of the weak, or those less able to survive, his pantheist God was there "all the time, waiting for His creation to discover Him."[233] Colling repeatedly makes the irresponsible mistake of comparing Darwinian evolution to the trial and error process used by our well-designed and very complex adaptive immune system. Unlike evolution, the immune system has a goal and a purpose: to fight off invaders by a preprogrammed, irreducibly complex mechanism that involves some random variation but filters out what it needs. Darwinian evolution is blind. It has no goal, no foresight, and no purpose except survival.

Colling never attempted to prove Darwinism. He simply assumed that it is true and succumbed to omnipresent phrases found in the writings of Darwin apologists, such as that evolution is supported by "overwhelming evidence" from the many sub-disciplines of biology. Relying on this, and other often-repeated claims are, for him, a valid substitute for evidence. A major motive for developing his view of origins clearly stems from what he considers the "embarrassment" that the "belief in an active creator God" has caused Christians.

Wanting to be socially acceptable and not an evangelical Christian outcast, he threw his lot in with the Darwinists. Becoming one of them, he now dogmatically concludes that "no serious scientific alternatives [to evolution] have surfaced—none! Evolution is the only current and viable scientific framework that provides a rational understanding of the immense and beautiful diversity of life on our planet."[234] He stated not only that "no one can ac-

---

230 Begley 2004.
231 Colling, 2004, p. 181.
232 Colling, 2004, p. 147, emphasis added.
233 Colling, 2004. Quoted in the dedication page of this book.
234 Colling, 2004, p. 16.

Tony added that he wrestles with why secularists like his son Bart choose to "live as though there is nothing and no one behind the awesome wonder of the universe or our most transcendent experiences, and then cherry-pick the theories and arguments that best support that lifestyle, [which] is an act of faith as well."[224] Bart chose this worldview as a result of his secular education and his lack of education in the opposing view.

Bart also relates that his hero, Ms. Ursula, the daughter of a minister, wrote that the scientific account of creation is the epic of evolution, which is "The Big Bang, the formation of stars and planets, the origin and evolution of life on this planet, the advent of human consciousness and the resultant evolution of cultures—this is the story, the one story, that has the potential to unite us, because it happens to be true."[225]

In short, Tony Campolo could not defend his faith, and a major reason was because he accepted evolution as God's means of creation, which seriously handicapped his ability to effectively defend his worldview. His son, Bart saw that once one accepts evolution, it is very difficult to defend the main reason most people give for believing in God, namely the argument from design.[226]

References

Campolo, Tony and Bart Campolo. 2017. Why I Left, Why I Stayed: Conversations on Christianity Between an Evangelical Father and His Humanist Son. New York: HarperOne.

Hailes, Sam. 2017. "Bart Campolo says progressive Christians turn into atheists. Maybe he's right." Premier Christianity. September 25. https://www.premierchristianity.com/Blog/Bart-Campolo-says-progressive-Christians-turn-into-atheists.-Maybe-he-s-right

---

[224] Campolo and Campolo, 2017, p. 139.
[225] Campolo and Campolo, 2017, p. 70.
[226] Campolo and Campolo, 2017.

Chapter 12

Richard Colling, Ph.D.

From YEC to Believer in a Mysterious Random Designer

Former Olivet Nazarene College Professor and former Chair of the Department of Biology, Dr. Richard Colling, although reared as a YEC, now argues that God is a mere "Random Designer," i.e., he works by natural selection via the agency of genetic damage called mutations. He concluded that God originally set up all of the basic laws of nature, and then allowed these laws to evolve the entire universe, including life, on its own. This position is effectively the same as deism, a worldview that is only a hair's breadth away from atheism.

Colling recognizes that the term "Random Designer … sounds like an oxymoron—a contradiction of terms" which it is—yet without embarrassment he openly advocates this view, one that is in direct contradiction to both Biblical and historical Christianity.[227] Colling teaches that God foreknew that randomness would eventually "accomplish His goals," whatever they were, from chaos to intelligently designed creation. Except in a very vague way, I could not find how Colling defines how this goal was or even could be achieved.

In short, he believes that Darwinian evolution, including random mutations - the main source of genetic variety - has generated all life on Earth, including humans. Somehow, God was mystically behind this random process, although how, where, and when is never mentioned by Colling in his book.[228] Of course, if God's intelligence was behind this process, it could not be random. Colling concedes that "Perhaps the Random Designer intentionally" guided what fully appears to be an unguided process, ironically making him out be an Intelligent Design supporter, a worldview that he adamantly opposes.[229] Begley writes that:

> Colling reserves some of his sharpest barbs for intelligent design, the idea that the intricate structures and processes in the living world—from exquisitely engineered flagella that propel bacteria to

---

227 Colling, 2004. p. 3.
228 Colling, 2004.
229 Colling, 2004, p. 147.

the marvels of the human immune system—can't be the work of random chance and natural selection. Intelligent design advocates look at these sophisticated components of living things, [and] can't imagine how evolution could have produced them and conclude that only God could have. That makes Colling see red.[230]

He concluded that "overwhelming evidence" exists that God is "truly a Random Designer."[231] His writings indicate that Colling might be closer to an agnostic, such as when he wrote "in spite of our inability to see, touch, or understand God in the physical dimension, we intuitively sense He is real—or at the very least, that He might be real."[232]

When all life was evolving by elimination of the weak, or those less able to survive, his pantheist God was there "all the time, waiting for His creation to discover Him."[233] Colling repeatedly makes the irresponsible mistake of comparing Darwinian evolution to the trial and error process used by our well-designed and very complex adaptive immune system. Unlike evolution, the immune system has a goal and a purpose: to fight off invaders by a preprogrammed, irreducibly complex mechanism that involves some random variation but filters out what it needs. Darwinian evolution is blind. It has no goal, no foresight, and no purpose except survival.

Colling never attempted to prove Darwinism. He simply assumed that it is true and succumbed to omnipresent phrases found in the writings of Darwin apologists, such as that evolution is supported by "overwhelming evidence" from the many sub-disciplines of biology. Relying on this, and other often-repeated claims are, for him, a valid substitute for evidence. A major motive for developing his view of origins clearly stems from what he considers the "embarrassment" that the "belief in an active creator God" has caused Christians.

Wanting to be socially acceptable and not an evangelical Christian outcast, he threw his lot in with the Darwinists. Becoming one of them, he now dogmatically concludes that "no serious scientific alternatives [to evolution] have surfaced—none! Evolution is the only current and viable scientific framework that provides a rational understanding of the immense and beautiful diversity of life on our planet."[234] He stated not only that "no one can ac-

---

230 Begley 2004.
231 Colling, 2004, p. 181.
232 Colling, 2004, p. 147, emphasis added.
233 Colling, 2004. Quoted in the dedication page of this book.
234 Colling, 2004, p. 16.

curately say that science supports" the creation or Intelligent Design position, but stresses "Let us be very clear: it doesn't."[235]

Colling has even rejected the central Christian doctrine of the fall and atonement, concluding that "Biblical accounts that proclaim the creation of the first man to be from the actual physical dust of the earth are … dubious."[236] Of course, this view is essentially that which is espoused by agnostics and atheists. He implies that the Adam and Eve creation account is symbolic, but what the account is symbolic of, he never says.[237] Colling, as Giberson (see chapter 17) also does, admits that evolution has challenged —and destroyed—almost every childhood belief that he once held about the Bible:

> As a young child, the Bible stories my parents often read to me relating the power and miracles of God captivated my imagination. … They related to me that God created the world and all living things in six days, molding Adam from the dust of the earth and Eve from Adam's rib. God was all-powerful and the Biblical story of creation explained how all life began. This view of creation was adequate for me until high school biology class, where I quickly learned a different view—the concept of evolution.[238]

As an adult, Colling attended a "Christian" college where he evidently lost the rest of his childhood faith. After learning about evolution, he no longer believed the basic teachings of Christianity, and even tried to explain the origin of life by abiogenesis:

> Two different explanations might account for the appearance of these early cells: A purposeful Designer—a God who instantaneously and supernaturally called … living cells into existence— or a process of random synthesis and selection which, given adequate time and appropriate conditions, … create the first life.[239]

He eventually concluded that the origin of life was not due to Intelligent Design, but rather was a result of "the random assembly of specialized biomolecules, followed by the preferential selection of the most valuable variants," which sparked "the whole sequence of creative events ultimately leading to

---

235 Colling, 2004, p. 16.
236 Colling, 2004, p. 113.
237 Colling, 2004, p. 113.
238 Colling, 2004, p. 5.
239 Colling, 2004, p. 44.

the first living cell on the planet—a very early rendition of random design."[240] This first cell eventually evolved into some multi-cellular life form, then into worms, fish, amphibians, reptiles, mammals, primates and, last, humans. He writes that, even though "pain and suffering" are a central part of evolution, this evil is balanced by evolution's many amazing achievements. For example

> the formation of consciousness and conscience in humans is ... a magnificent and monumental triumph! Beginning with the simplest biochemical reactions and building upon the resulting structures and assemblies, the Random Designer has brought forth a most spectacular creation.[241]

Colling has even uncritically accepted some of the uninformed myths about Darwinism critics often repeated by atheists, such as non-Darwinists have "not engaged in any original peer-reviewed scientific research, nor offered any viable scientific alternatives" to evolution.[242] Obviously, he is totally unaware of the vast literature produced by both Intelligent Design theorists and creationists.

Colling openly recognizes the enormous complexity of life, even noting that "while explaining some fine detail regarding the human cell, I have faced a lecture hall filled with university biology students and literally felt chills run down my spine as I was gripped by the unbelievable beauty and order so evident in life." In his book, Colling accurately explained the basic workings of DNA, the immune system, and other aspects of cellular biology at a high school level. But he argues that God had no detectable role in creating this "unbelievable beauty and order"—rather that the Random Designer (evolution, meaning natural selection of genetic damage) did it all.[243]

If an ID-proponent wrote half the Designer-praising evangelical statements that Colling did, it would be widely repeated as proof that ID is religion! One looks in vain for a clear exposition of his theology, but instead one finds flowery prose praising the wonders of the "Random Designer" who constructed all life, one step at a time, from nothing more than damage to DNA that causes what we call mutations, plus natural law, time, and natural selection.[244]

---

240 Colling, 2004, pp. 44-45.
241 Colling, 2004, p. 117.
242 Colling, 2004, p. 6.
243 Colling, 2004, p. 7.
244 Colling, 2004, p. 61.

After irresponsibly claiming that an "infinite number of possible protein structures provide an infinite number of possible protein functions, an infinite number of possible cell types, and an infinite number of different life forms," he asserts that the natural selection of mutated DNA is an

> incredibly productive method for creating diversity [and] is another example of random design. ... I must confess, these processes never cease to amaze me. But while we may find them amazing, the Random Designer is not the least bit astonished. Unlimited potential is His very nature, and random design is part of His process and plan. He beckons us to come close and learn more.[245]

Colling recognized that "the explosive conflict surrounding evolution in our culture today is not really about the science at all. It arises from two conflicting worldviews: the atheistic worldview versus a fundamental creationist worldview."[246] Posing this false dichotomy, he sides with the atheists in the academy, concluding that God is "more concerned with final outcomes than with intermediate processes, nonproductive pathways ... In a very real sense then, failure abounds" in the evolution process.[247]

How he could know all of this about a Creator that he is not sure even exists is left unstated, but it hardly conforms to the creation account as taught by the Christian church for the last 2,000 years and the Jews for 2,000 years before that. His ideas, which not unexpectedly caused quite a stir at Olivet Nazarene University where Colling was a professor, are far closer to New-Age teachings than to the basic Judeo-Christian theology tradition.

In short, his writings give no viable reason to believe in God and plenty of reasons not to. A professor who wrote that God "might be real" would likely be of little help to college students struggling with their Christian faith as, in my experience, many students are.[248] We agree with his statement that "the primary goal of science is simple—to learn how things work."[249] If the science establishment and Colling would stick to that goal, and not invent worldviews in an attempt to harmonize Darwinism with its opposite, Christianity, his science would cause far fewer problems in society and the church. Professor Colling is right about one thing: his Random Designer is not a de-

---

245 Colling, 2004, p. 55.
246 Colling, 2004, p. 15.
247 Colling, 2004, p. 72.
248 Colling, 2004, p. 147.
249 Colling, 2004, p. 11.

signer in any sense of the word.[250]

References

Begley, Sharon. *2004.* "God Made Evolution, Prof Says." *The Journal Gazette*, Saturday, December 18, p. 2C.

Colling, Richard. 2004. Random Designer: Created from Chaos to Connect with the Creator. Bourbonnais, IL: Browning Press.

Audio: Random Design: Building Bridges Between Science and Faith (Part 1) (10:00) https://www.youtube.com/watch?v=KfHuCWFxWYI

---

250 Colling, 2004, p. 3.

Chapter 13

Francis Crick, Ph.D.

The Rock Star of Modern Genetics Becomes an Atheist

Some 'Rock Stars' of 20th century science have abandoned whatever religious faith they were taught when their child-like reasoning went unchallenged. For example, Nobel Prize scientist Francis Crick recounts, as a very young man he began to suspect that his religious views were wrong, and "from then on I was a skeptic, an agnostic with a strong inclination toward atheism."[251] The churches he knew were no match for the lure of perceived wisdom in academia. The reasons why he rebelled against Christianity, he says, were his "growing interest in science and the rather lowly intellectual level of the preacher and his congregation."[252]

He added that his "loss of faith in Christian religion, and my growing attachment to science have played a dominant part in my scientific career ... in the choice of what I have considered interesting and important."[253] He became focused on proving evolution and disproving creationism of all stripes. In fact, during the 50th anniversary year celebrating his discovery of the double helix he commented that "The God hypothesis is rather discredited." Indeed, he says his "distaste for religion was one of his prime motives in the work that led to the sensational 1953 discovery."[254]

James Watson, with whom Crick shared the Nobel Prize, admits a similar boyhood experience that ended in his rejection of theism. In the year 1940, when he was twelve, he was confirmed in the Church, but soon after he concluded that because

> religious explanations seemed to be "myths from the past," he gave up mass to go birdwatching: "I came to the conclusion that the Church was just a bunch of fascists that supported Franco. I stopped going on Sunday mornings and watched the birds with my father instead."[255]

One could add that his experience was also due to the lack of interest in

---

251 Crick, 1988, p. 11.
252 Crick, 1988, p. 10.
253 Crick, 1988, p. 11.
254 Highfield, 2003.
255 Watson, quoted in McElheny, 2004. p. 9.

creation in the churches he was involved in. Crick has never made the common claim that religion was shoved down his throat, but rather only that his "parents were religious in a rather quiet way."[256] He added that, at about age twelve, when he told his mother that he no longer wished to attend church, "she was visibly upset."[257]

Crick's aforementioned "loss of faith in the Christian religion and ... growing attachment to science ... played a dominant part in [his] scientific career,"[258] bringing him to realize early in life that a "knowledge of the true age of the earth and of the fossil record makes it impossible for any balanced intellect to believe" in Genesis.[259]

A major point he stresses is "if some of the Bible is manifestly wrong," such as its account of creation, "why should any of the rest of it be accepted?"[260] Crick concluded, "what could be more foolish than to base one's entire view of life on ideas that, however plausible at the time, now appear to be quite erroneous?"[261] A major life goal of his was to understand some of the mysteries of the universe that remain unexplained because these areas still "serve as an easy refuge for religious superstition."[262]

He spent much of the rest of his career endeavoring to "explain" these refuges for religion—such as the origin of consciousness—from a naturalistic perspective. In this goal, he was spectacularly unsuccessful. Even many of his fellow atheists felt that his time would have been much better spent in the area of molecular biology where he did his important work instead of trying to explain "these refuges for religion."

Crick acknowledges the enormous complexity of living things which impressed our forefathers, who considered that it is

> inconceivable that such intricate and well-organized mechanisms would have arisen without a designer. Had I been living 150 years ago I feel sure I would have been compelled to agree with this Argument from Design. A most thorough and eloquent protagonist was the Reverend William Paley whose book, Natural Theology ... was published in 1802."[263]

---

256 Crick, 1988, p. 10.
257 Crick, 1988, p. 10.
258 Crick, 1988, p. 11.
259 Crick, 1988, p. 11.
260 Crick, 1988, p. 11.
261 Crick, 1988, p. 11.
262 Crick, 1988, p. 11.
263 Crick, 1988, pp. 24–25.

After stating that this once "compelling argument was shattered by Charles Darwin, who believed that the appearance of design is due to the process of natural selection," Crick fell into the belief of other atheists, like Richard Dawkins, who claim that design is an illusion—the natural living world appears to be designed but is, in fact, not designed but rather is a result of natural selection acting on gene mutations.[264]

He concluded that those who reject Darwinism are either "not aware" of, or are "indifferent" to, the fact of Darwinism. He never could accept the views of the many well-qualified scientists and intelligent laypeople who, not for religious reasons, have assessed Darwinism and rejected it on the basis of solid scientific evidence and logic.[265]

References

Crick, Francis. 1988. *What Mad Pursuit: A Personal View of Scientific Discovery*. New York: Basic Books.

Highfield, Roger. 2003. "Do our genes reveal the hand of God?" *The Telegraph*, March 20.

McElheny Victor K. 2004. Watson and DNA: Making A Scientific Revolution. New York: Basic Books.

Video: *Interview with James Watson and Francis Crick* (38:57) https://www.youtube.com/watch?v=NGBDFq5Kaw0

Audio: *DNA Nobel Prize Winner Francis Crick Says our genes Were Planted Here By Aliens* Directed Panspermia (3:05) https://www.youtube.com/watch?v=8XFUyafUmso

---

[264] Nowhere in his writing does he indicate the fact that natural selection can account only for the survival- of-the-fittest, not the arrival-of-the-fittest.
[265] He naively says, "in Western society a rather vocal minority are actively hostile to evolutionary ideas" (Crick, 1988, p. 25). At least in the USA he was the vocal minority: the latest poll shows more than half of all Americans disagree with evolutionary naturalism. Crick then elaborates the traditional gradualist view espoused by Darwin, ignoring what some scientists conclude are lethal criticisms of this view.

Chapter 14

Raymond Dart, Ph.D.

From Devout Christian to Freethinker

Raymond Dart was a fossil hunter looking for the missing link in an attempt to prove human evolution. Dart, an Australian anatomist and anthropologist, is best known for his involvement in the discovery of the first fossil fragments found of Australopithecus africanus, that evolutionists claim was an extinct hominid closely related to humans. The fossil was found at Taung in the northern part of South Africa, on the edge of the Kalahari Desert in 1924, thus the fossil is called the Taung Child.

His father, Samuel, a Baptist, was elected General Visiting Superintendent of the Methodist Sunday Schools circuit. He faithfully reared his family in all the Christian precepts of the New Testament. Raymond was baptized in the new local German Baptist Church and received into membership there in 1907. Concerning his youthful Christian beliefs, Dart wrote he was reared "in a devout Methodist and Baptist family environment,"[266] gladly sharing also the fundamentalist philosophy of his conservative Plymouth Brethren family and friends.

Raymond intended to become a medical missionary in China, an ambition that he often expressed to his fellow students during his school years. As an adult, Raymond always carried two Bibles around with him, one in English, and the other in German. He could, if challenged, recite chapter and verse from Scripture in either language.

Raymond's youthful Christian faith and his strong goal to become a medical missionary soon succumbed to the theory of evolution in college. Dart wrote, "My first frank confrontation with evolutionary ideas was in 1911 as a biology student in the University of Queensland."[267] Then, he says that, in 1914, he fell under the spell of Grafton Elliot Smith, who worked at the University of Manchester and was one of the world's preeminent neuro-anatomists, and a fellow Australian who was later knighted. Smith is remembered today for giving brilliant public lectures on the evolution of the human brain. This is ironic in that he strongly supported the validity of the Piltdown man, allegedly discovered in 1910-1912, which turned out to be a human skull

---

266 Harold Dart. 1981, p. 110.
267 Tobias. 1984, p. 2.

joined to an ape jaw, and not a creature in-between an ape and a human as claimed for decades until proven to be a hoax in 1953.

Dart also accepted the views of Professor J. T. Wilson at Sydney University who, he wrote, had "a deep respect for the Darwinian and Haeckelian process of utilizing vestigial comparative anatomical structures as clues to reveal our extremely ancient prehistory (i.e. phylogeny)."[268] Dart (and no doubt also Professor Wilson) presumably was unaware that the 'evidence' put forward by Haeckel was also totally fraudulent. At the age of 29, Dart had drifted so far from his childhood faith that he had described himself in his application to the University of the Witwatersrand as a 'Freethinker.' In 1922 Dart became Professor of Anatomy at the newly-founded University of Witwatersrand in Johannesburg, South Africa.

References

Dart, Harold. 1981. Happenings: Historic, heroic, and hereditary: a saga of events prior to, during, and subsequent to the arrival of the first of the author's relatives in Australia on 27th February, 1837. New South Wales Publication.

Tobias, Phillip V. 1984. Dart, Taung, and the 'Missing Link.' Johannesburg, South Africa: Witwatersrand University Press.

---

268 Tobias. 1984, p. 2.

Chapter 15

Richard Dawkins, Ph.D.

From Creationist to Militant Atheist

Oxford University Professor Richard Dawkins, probably the most famous atheist living today, as a youth joined the Church of England, "and for some time carried 'a guttering torch' for it."[269] He was "persuaded by the ... argument from design" to believe in God until he "discovered Darwinism," and soon the last vestige of his "religious faith disappeared forever."[270] He obtained his doctorate in Zoology from Oxford University, determined to destroy both religion and God. In his book The God Delusion, he wrote; "If this book works as I intend, religious readers who open it will be atheists when they put it down."[271] He now travels throughout the world speaking to audiences of thousands of people and sells multi-millions of books in dozens of languages spreading his message of unmitigated hatred toward the God of the Bible.[272]

Seeds of Doubt Planted

The first seeds of doubt about Christianity were planted in Dawkins at the age of about nine when he learned from his mother that Christianity was only one of many religions that sometimes contradicted each other. He reasoned that "they couldn't all be right, so why believe the one in which by sheer accident of birth, I happened to be brought up." He added, "after my brief phase of going to Communion, I gave up believing in everything that was particular about Christianity, and even became quite contemptuous of all particular religions."[273]

Nonetheless, at this time in his life, Dawkins retained "a strong belief in some sort of unspecified creator, almost entirely because I was impressed

---

269 Blanchard, 2010, p. 7.
270 Blanchard, 2010, p. 7.
271 Dawkins, 2006, p. 5.
272 For many examples see his book The God Delusion. New York: Mariner Books. 2008. Other publications which reference his ideas include God Is Not Great: How Religion Poisons Everything by Christopher Hitchens. New York: Twelve Publishers. 2009; and The End of Faith: Religion, Terror, and the Future of Reason. By Sam Harris 2005. New York Norton, 2005 and Fighting Words: The Origins of Religious Violence *by Hector Avalos Amherst, NY: Prometheus Books. 2005.*
273 Dawkins, 2013, p. 140.

by the beauty and apparent design of the living world, and—like so many others—I bamboozled myself into believing that the appearance of design demanded a designer."[274]

He added that studying the world and nature had convinced him that "there must have been a designer." Ironically, at this time his faith was reinforced by the American rock star Elvis Presley, of whom he was an enthusiastic fan. He "worshipped Elvis" and, at this time was "a strong believer in a non-denominational creator god."[275] Once, when Dawkins passed by a shop window, he spotted an Elvis album titled Peace in the Valley that featured the song "I Believe." He says that he was transfixed and listened to the song with delight, noting that every time his hero sang about

> the wonders of the natural world around him, he felt his religious faith reinforced. My own sentiments exactly! This was surely a sign from heaven … calling me to devote my life to telling people about the creator god - which I should be especially well qualified to do if I became a biologist like my father. This seemed to be my vocation.[276]

He added that at that time, he had not yet "worked out the elementary fallacy of" the design

> argument, which is that any god capable of designing the universe would have needed a fair bit of designing himself. If you are going to allow yourself to conjure a designer out of thin air, why not apply the same indulgence to that which he is supposed to have designed … of course, Darwin provided the magnificently powerful alternative to biological design which we now know to be true. Darwin's explanation had the huge advantage of starting from primeval simplicity and working up, by slow, gradual degrees, to the stunning complexity that pervades every living body.[277]

Dawkins concluded that he is not proud today of the period of religious frenzy that he experienced as a youth, and is today glad to admit that he

> became increasingly aware that Darwinian evolution was a power-

---

274 Dawkins, 2013, pp. 140-141.
275 Dawkins, 2013, pp.141-142.
276 Dawkins, 2013, pp.141-142.
277 Dawkins, 2013, pp. 140-141.

fully available alternative to my creator god as an explanation of the beauty and apparent design of life ... although I understood the principle [of evolution], I didn't think it was a big enough theory to do the job. I was biased against it by reading Bernard Shaw's preface to Back to Methuselah.[278]

The reason was, Shaw favored Lamarckian evolution because it was purpose-driven and he

hated Darwinian (more mechanistic) evolution, and I was swayed [and]...went through a period of doubting the power of natural selection to do the job required of it. But eventually a friend ... persuaded me of the full force of Darwin's brilliant idea and I shed my last vestige of theistic credulity, probably at the age of about sixteen. It wasn't long then before I became strongly and militantly atheistic.[279]

He writes that today Charles Darwin is his "greatest scientific hero. Philosophers are fond of saying that all philosophy is a series of footnotes to Plato," adding that a

far better case could be made that all of modern biology is a series of footnotes to Darwin. And that would be a genuine compliment to the science of biology. Every biologist treads in Darwin's footsteps and, in all humility, none of us could do better than to follow his example.[280]

A key influence on Dawkins' embrace of Darwinism was a tutor who shaped his ideas on science: the zoologist Dr. John Currey, later Professor of Zoology at York University, who revealed to him that

his—and now my—favorite example of revealingly bad 'design' in animals; the recurrent laryngeal nerve ... instead of going directly from the brain to its end organ the larynx, this nerve makes a detour ... down the chest, where it loops around a large artery before proceeding back up the neck to the larynx. This is eloquent of terribly bad design, but is completely explicable the moment

---

278 Dawkins, 2013, p. 142.
279 Dawkins, 2013, p. 142.
280 Dawkins, 2013, p. 290.

you forget design and start thinking in terms of evolutionary history instead. In our fishy ancestors, the shortest route for the nerve was posterior to the then equivalent of that artery which... supplied one of the gills. Fish don't have necks. When necks started to lengthen on land, the artery gradually moved backwards relative to the head, step by tiny step through evolutionary time further away from brain and larynx. The nerve kept abreast...as evolution progressed [making] a longer and longer detour until, in a modern giraffe, its diverted route is a matter of several meters.[281]

This so-called flaw in design, has now been fully refuted, a fact of which Dawkins is obviously not aware.[282] Dawkins' horrific caricature of who God is was caught on camera as he was interviewed by Ben Stein for the movie "Expelled."[283] The same interview where he [Dawkins] was forced to admit that, though he vehemently objected to the Judaeo-Christian God, he could accept a distant, extra-terrestrial first-cause being responsible for seeding life on Earth (i.e., an evolved intelligent-designer of a lesser sort). Amazing!

References

Bergman, Jerry. 2019. "The Vas Deferens is not Poor Design." In *The "Poor Design" Argument Against Intelligent Design Falsified*. Tulsa, Oklahoma: Bartlett Publishing.

Blanchard, John. 2010. *Dealing with Dawkins*. London: Evangelical Press.

Dawkins, Richard. 2006. *The God Delusion*. Boston: Houghton Mifflin

Dawkins, Richard. 2013. *An Appetite for Wonder: The Making of a Scientist*. New York: Ecco (an imprint of HarperCollins).

Frankowski, Nathan, Director. 2008. *Expelled: No Intelligence Allowed*. Dawkins was interviewed by Ben Stein in this documentary. https://www.youtube.com/watch?v=YmAFp27gYFA&t=14s

---

281 Dawkins, 2013, pp. 159-160.
282 Jerry Bergman. 2019. Also see the upcoming book by Jerry Bergman: The Poor Design Argument Refuted, (in press).
283 Frankowski, 2008

Videos

Video: Richard Dawkins Interviews Creationist Wendy Wright (1:06:42)
https://www.youtube.com/watch?v=-AS6rQtiEh8
This video is a gem!

Video: Richard Dawkins Teaching Evolution to Religious Students (52:26)
https://www.youtube.com/watch?v=jNhtbmXzIaM

Audiobook: The God Delusion. Read by Richard Dawkins (11:58:37)
https://www.youtube.com/watch?v=lgKLmxIu1y4

Video: Alan Lightman & Richard Dawkins on Science & Religion (1:06:12)
https://www.youtube.com/watch?v=eSCDfjTDVCk

Chapter 16

Eugene Dubois, Ph.D.

From God Believer to Evolution Believer

Eugene Dubois was born into a conservative Catholic Dutch family on January 28, 1858, about a year before Darwin published his Origin of Species in 1859.[284] Dubois' strong interest in science motivated his father to send him to the state technical high school, a school with laboratories of similar quality found in a small university. Several local Catholics objected to Eugene attending this school because they feared—correctly as it turned out—that he would learn their ideas including Darwin's new evolutionary theory.[285] Shipman added that the village elders were horrified that

> Dubois would consider sending his son to such a place. "He'll lose his religion," they predict, nodding their heads with conviction.... "They'll teach him all those anti-Christian theories, and soon he'll believe them. He's a nice boy, a smart boy, but the mayor will be sorry if he sends his son to such a place!" In the end, they ... fuss[ed] so tiresomely that Jean Dubois decided to send Eugéne to the HBS in part to defy them. ...By the end of his first year at Roermond, when he is thirteen, he is starting to question the teachings of the Church. ... [and] begins to doubt everything, almost reflexively.[286]

After he graduated from the public state school at age 19, Dubois entered medical school where he was "exposed to the exciting ideas of Darwinian biologists."[287] His Dutch Catholic upbringing openly conflicted with what he was learning about evolution in school—and evolution soon won out.[288] Dubois lost his religion as his father's friends predicted, and he became not only a life-long Darwinist, but also an active opponent of Christianity.

While a student at Jena University, Dubois' major professor, the now infamous Ernst Haeckel, was an enthusiastic follower of Darwin. Darwin

---

284 Shipman, 2001, p. 11.
285 Shipman, 2001, p. 19.
286 Shipman, 2001, p. 19.
287 Milner, 1990, p. 147.
288 Regal, 2004, p. 64.

had predicted an evolutionary line of links existed between modern humans and their ape-like ancestors, but until the discovery of the Java Man bones, no plausible candidates were then known.[289] In 1877, only a "scrap or two of fossil evidence for human ancestors, notably the Neanderthal skullcap from Germany," existed.[290]

Dubois cultivated a strong drive to find scientific proof for Darwinism—specifically to find the "missing" link, the "Pithecanthropus," Greek for ape-man.[291] He knew that finding the missing link "would be the greatest scientific discovery ever."[292] Dubois' powerful motivation to find this missing link was to disprove theism in order to support his belief that "there is no truth in religion"—and he was drawn to prove evolution "with an almost religious fervor."[293]

Although trained as a physician and an anatomist, Dubois abandoned both his home and a promising career as a Professor at the University of Amsterdam to search for some fossil evidence of evolution. To achieve this goal, he took his young wife and small children halfway around the globe to search for Darwin's missing link in a remote part of Dutch East India now known as Indonesia. Because he had no concrete evidence that this island could produce any useful results, many evolutionists felt that his quest was based on only a foolish hunch that he would find the missing link there.

Dubois arrived at the idea of not journeying far away Java from several sources, including reading Alfred Russel Wallace's account of the human-like animal now called an orangutan that lived on the Sumatra islands.[294] Ernst Haeckel's book titled Theory of Asian Human Origins was also important in Dubois' conclusion that humans first evolved on, or near, Java.[295]

Haeckel was right about one thing: Java was a very good location to look for fossils, and he found literally tons of them in Java. He shipped 400 cases of the most interesting ones back to Holland, and his workers found so many bones that they sold large numbers to be ground up and sold for various medical nostrums.[296] The bones commonly found included various fish, reptiles, mammals (elephant, rhinoceros, hippopotamus, tapir, ruminants, mon-

---

289 Bowden, 1977.
290 Milner, 1990, p. 147.
291 Keith. 1925, p. 438.
292 Shipman, 2001, p. 22.
293 Shipman, 2001, pp. 19, 24.
294 Milner, 1990, p. 147.
295 Regal, 2004, p. 65.
296 Milner, 1990, p. 147.

key), and even mollusks of a type still living in the area.[297] The story of Java man turned out to be another Piltdown man type forgery.[298]

References

Boule, Marcellin and Henri Vallois. 1957. *Fossil Men*. New York: The Dryden Press.

Bowden, M. 1977. *Ape-Man—Fact or Fallacy: A Critical Examination of the Evidence*.

Bromley, KY: Sovereign Publications Keith, Sir Arthur. 1925. *The Antiquity of Man*. Covent Garden, WC, London: Williams and Norgate.

Keith, Sir Arthur. 1925. *The Antiquity of Man*. Covent Garden, WC, London: Williams and Norgate.

Milner, Richard. 1990. *The Encyclopedia of Evolution*. New York: Facts on File.

Regal, Brian. 2004. *Human Evolution: A Guide to the Debates*. Santa Barbara, CA: ABC Clio.

Shipman, Pat. 2001. *The Man Who Found the Missing Link: Eugene Dubois and his Life Long Quest to Prove Darwin Right*. NY: Simon and Schuster.

---

297 Boule and Vallois, 1957, p. 113.
298 For a detailed discussion in this man and his life see chapter 15 in Jerry Bergman's *Darwinism's Frauds, Blunders and Forgeries*. Atlanta, GA: CMI Publishing, 2017.

Chapter 17

David Duke

From YEC to Evolutionist to Head of the KKK

In his autobiography, David Duke, was a former leader of several racist groups including the Ku Klux Klan and the American Nazi Party. According to Zatarain, Duke has "become a political rock star of sorts"—and was one of the most well-known Americans of the 1970s to around the year 2000.[299] Furthermore, Duke has worked with virtually every prominent American racist active during the last 40 years.[300] Duke's popularity can be gauged by the fact that he received 680,000 votes in the 1991 Louisiana gubernatorial runoff, and was elected to serve in congress for the State of Louisiana.[301]

His Religious Background

Duke was reared a Methodist (his father was a Sunday school teacher), and later attended the Church of Christ.[302] Duke covers his early religious upbringing in his autobiography, which details why he rejected Genesis and creationism, as well as the single origin of all human races from Adam. In order to learn "how racial differences originated," he studied evolutionary theory in detail.[303] In short, when he accepted Darwinism, he rejected both the Bible and his church's teaching, and instead relied on evolution to form his conclusions on race, which he concluded was an overwhelmingly proven scientific fact.

Once he understood the Darwinist teaching on "the realities of racial difference," he realized that "by learning about the evolutionary forces that created the different races, we can understand the character ... of the various races, our own included."[304] He noted that many Darwinists used the terms "race" and "breed" interchangeably, and so he applied the research on the evolution of animal "breeds" to humans.

The conflicts Duke had with the church were not only about Darwin-

---

299 See Zatarain, 1990, p. 10 for the rock star quote.
300 Bridges, 1994, pp. 41, 115.
301 Bridges, 1994, p. 2.
302 Duke, 1998, p. 256.
303 Duke, 1998, p. 89.
304 Duke, 1998, p. 90.

ism, but also with the church's teaching of Genesis. He bemoaned the fact that, when he graduated from college in the mid-1970s, an increasing number of churches were teaching that racism was a sin.[305]

His Religious Battle

Duke's father, a geologist, tried to reconcile evolution with Christianity by surmising that evolution might have been the means that God used to create life. This background set the groundwork for Duke's later acceptance of Darwinism. From a young age, Duke regularly read about science in Science Digest, National Geographic, and other science magazines.[306] As he read more and more on "the scientific issue of race," he became torn between his religion and science."[307]

He became involved in researching Darwinism while still attending a Church of Christ school in New Orleans. As a result of his study of evolution, Duke openly challenged his Sunday school teachers by discussing his evolving ideas about human origins, and their implication for racism. When endeavoring to combine his Darwinist racist beliefs with Christianity, he attempted to rationalize the plain statements of the Genesis creation account, using many of the same rationalizations used by theistic evolutionists today.

Duke eventually sided with Darwinism and rejected creationism. He concluded that, with "each passing day more evidence emerges of the dynamic, genetically-born, physical and physiological differences between the races."[308] This belief also ended his commitment to orthodox Christianity.[309] After his acceptance of Darwinism, Duke firmly believed that "all life on Earth had evolved and is still undergoing change."[310] He then presumed that both the European and Asian races were at a "higher level of human evolution than the African race."[311]

Especially important in Duke's conversion to Darwinism was the "hard evidence of the great age of the Earth—such as the eras of geological time it took to raise Mount Everest from the bottom of the sea."[312] This supposed evidence caused Duke to reject the Biblical account of Creation (even

---

305 Duke, 1998, p. 257.
306 *Duke, 1998, p. 21.*
307 Zatarain, 1990, p. 80.
308 Duke, 1998, p. 103.
309 Bridges, 1994, p. 7.
310 Duke, 1998, p. 101.
311 Duke, 1998, p. 103.
312 Duke, 1998, p. 103.

broadly interpreted), and to accept the Darwinist interpretation. Long ages also figure prominently in Duke's racist arguments. He concluded that the amount of time Darwinists believe that blacks and whites have been separated by evolution is more than enough time to produce what he views as the profound differences that exist between the human races.[313]

Duke also argued that "denying the reality of race is a good example of how race "egalitarians" are grasping for straws. He alleged that, controlling for the environment, racial divergence by evolution had produced approximately one standard deviation difference in I.Q. between modern blacks and whites.[314] This difference would require an increase of only a tiny fraction of one percent (.003) in I.Q. during each generation of whites. This belief relied on works such as Elmer Pendell, (referenced later in this chapter),which reviewed the now debunked research on I.Q. and race, to conclude that heredity plays "a leading role in intellectual ability."[315]

Duke derisively called the "creationist belief that God instantaneously created mankind and all of Nature ... egalitarianism," and complained that egalitarianism became the "dogma of our times." He was especially critical of creationism because creationists were race egalitarians who teach that "God made us all the same." He bemoaned the fact that the mass media has helped to convert "both the scientific community—which espoused evolution and fundamentally opposed creationist community—into spouting almost an identical egalitarian dogma."[316] Duke claimed that "anyone in the religious community who dared to tell the truth of race [by which he meant black inferiority] was accused of being against God Himself."[317]

Duke used not only Darwinist arguments to justify racism, but also quoted Scripture. For example, he quoted the Scriptures, which he took out of context, that stated slaves (some translations used the word servants) should be obedient to their earthly masters (Ephesians 6:5, Timothy 6:2, and Titus, 2:9-11). In biblical days, slavery was not race based. A slave in Rome could become free by merit and work. Some were able to buy their freedom, and some even became kings or high government officials.

Integral to Duke's racism is the supposition that Darwinism is science. His beliefs in this area are similar to those of the early leaders of eugenics (all Darwinian evolutionists), who had led America down the path of scientific

---

313 Duke, 1998, pp. 90-91.
314 Duke, 1998, p. 87.
315 Duke, 1998, p. 188.
316 Duke, 1998, pp. 102-103.
317 Duke, 1998, p. 103.

racism in the early 20th century. Starting with Darwin and his half-cousin Galton, such beliefs fed directly into the Nazi genocides and all the other tragic consequences so deplored today. His conclusions in this area are similar to those of the early eugenic leaders who had an important role in American history at the turn of the last century, and also in Germany during Nazi rule. Duke discussed in some detail both positive and negative eugenics, implying support for both.

A concern repeatedly discussed by Duke is dysgenics—race degeneration. He strongly believed that pure or 'favored races' might become weakened by Caucasians interbreeding with the "inferior" black races. Duke makes clear in his autobiography that his racism is clearly a result of his acceptance, not only of Darwinism, but of the eugenics that logically results from Darwinism. Duke also repeated many of the arguments that were commonly published in the mainline biological literature until the end of World War II and the American civil rights movement – such as claiming that evolution produces racial differences not only in skin color and hair texture, but also in brain size differences, cranial structure, intelligence, musculature, hormonal levels, sexual behavior, temperament, dentition, and even personality.[318]

Duke Confronts the Critics of Racism

Duke also reviewed the various scientific arguments against racism, such as Ashley Montagu's Man's Most Dangerous Myth: The Fallacy of Race. He decided that Montagu's "myth of race" argument is analogous to saying that dog breeds are a myth because one can find specific traits that exist in various breeds: "I thought about the question long and hard, and I asked myself, 'because some similar traits are found in different breeds of dogs, does that mean there are no St. Bernard's or Chihuahuas?'"[319]

Duke also reviewed Jared Diamond's arguments against racism, which he tried to refute by noting that the "closest relatives to man are the recent primates who are also relatively close in DNA. Chimpanzees, for instance, share 98.5 percent of the DNA with people."[320] He then argued that this claim is invalidated by the claim that Black and White DNA differ by less than one percent. Duke reasoned that, since only a 1.5 percent difference in DNA between humans and chimpanzees produced humans with brains about twice as large as chimps, small DNA differences could produce large differ-

---

318 Duke, 1998, p. 86.
319 Duke, 1998, p. 85.
320 Duke, 1998, pp. 103-104.

ences in the human races.[321] Duke concludes that "If one follows Diamond's rationale, there is no difference between humans and chimpanzees because we can find sets of arbitrary selected genetic traits we share."[322]

The 98 percent claim is often used as an argument in favor of Darwinism, and is uncritically repeated by Duke, even though the exact difference between humans and chimpanzees cannot be determined until much more is known about both the human and chimpanzee genome, especially the function and structure of chimpanzee genes. Estimates now range to almost 88 percent similarity; thus a 320 million base pair difference exists.[323] Duke concludes that "the vast majority of the basic genes that make up the races are not only shared by them, but also by all mammals and even all other orders of life. The important distinctions are the small percentage of genes that affect the structure and composition of those life forms."[324]

After studying anthropological theories about the origin of races, Duke summarized the two dominant theories of evolution—the single-origin hypothesis, and the multi-regional hypothesis advocated by University of Michigan Anthropology Professor Milford Wolpoff. The single-origin theory argued that the different races crossed the Homo sapiens threshold separately during evolution. Duke was especially impressed with the research that postulated Homo sapiens first evolved in Africa and then "evolved separately into two distinct genetic groups, the African and the non-African, about 120,000 years ago."[325]

Duke's belief that the major races have been in existence for tens of thousands of years meant there was "more than enough time for geography and climate to have created [by evolution] the profound differences that exist" today between the races.[326] The Darwinist conclusion that the Caucasians and Negroid groups have been divided for at least a 110,000 year period convinced Duke that significant differences must exist between them.[327] In contrast, Caucasian and Asians have been separated by only forty thousand years.

For this reason, far fewer differences exist between Asians and Caucasians than exist between Negroids and Caucasians, who were separated long before this in the past. Duke repeatedly stressed that his conclusions on race were based on scientific research completed by leading evolutionists, and that

---

321 Duke, 1998, pp. 85-86.
322 Duke, 1998, p. 85.
323 Bergman and Tomkins, 2012; Tomkins and Bergman, 2012.
324 Duke, 1998, p. 86.
325 Duke, 1998, pp. 90-91.
326 Duke, 1998, p. 91.
327 Duke, 1998, p. 91.

this research forced him to reject the biblical Creation account he was reared to believe.[328]

Darwinists Who Influenced Duke

Sociobiology, as advocated by Harvard's Edward Wilson and other biologists, was also critically important in the devolvement of Duke's thinking. Especially critical was "the landmark work of Dr. Edward Wilson in his seminal Sociobiology: The New Synthesis." Duke read this work a few months after it was published and "found it magnificent."[329]

Duke admits that his interest in "the effects of evolution on race" was originally first stirred by Professor Carleton Coon, who was still an active scientist when Duke was doing his research. Coon's racist ideas were then mainline, and influenced hundreds of his students, who themselves became anthropology professors at several leading American universities. He was then at that time a leading physical anthropologist and the President of the American Association of Physical Anthropologists. Coon published his many books with major publishers and, at the time of his death, was a Research Associate at the Peabody Museum of Harvard University.

Duke read all of Coon's books he could find, including the *Living Races of Man, Story of Man, Origin of the Races,* and *The Races of Europe.*[330] Zatarain claimed that it was Coon who "introduced Duke to the view that race was a key factor in the development of modern man."[331] Duke was also heavily influenced by many other Darwinists, especially Harvard Professor of Anthropology, Earnest Hooton. Although Duke relied on many pre-1960 evolutionist writings in which racism was a dominant topic, he also quoted more recent Darwinists.

Professor Elmer Pendell's Influence

Another major influence on Duke was Professor Elmer Pendell's works, including *Sex Versus Civilization* (1967) and *Why Civilizations Self-Destruct* (1977). Both books concluded that more focus needs to be on the issue of human quality, as opposed to an almost exclusive focus on human

---

328 Maginnis, 1992.
329 Duke, 1998, p. 451.
330 Many editions of these and other books authored by Professor Coon exist. A search on Addall.com will bring up scores of copies.
331 Duke, 1990, p. 79.

quantity.³³² Dr. Pendell, the editor/author of a major textbook (1942), taught at Cornell and Penn State. He holds degrees from both Cornell and the University of Chicago.

From Pendell, Duke obtained the idea that the less intelligent and less fit, as a whole, reproduce faster than the most intelligent and most fit.³³³ Pendell's solution was to have the state regulate reproduction according to eugenics principles, which translates into sterilization of "inferior" humans.³³⁴ Professor Pendell stressed "the only source of brains is heredity," and the key to evolution is "the elimination" of the less fit.³³⁵ As a result, "as below average individuals were wiped out, the average moved up the scale ... the weeding-out aspect of biological evolution has worked in the human species as well as in other species" and that "the culling of human flocks was basic to the development of mentality."³³⁶ Pendell concluded that he was only "following through" on Darwin.³³⁷

Condemns Race Mixing

It is clear that Darwinism was at the heart of Duke's racist position. Many of his arguments come from leading mainline Darwin theorists—some from the pre-civil rights era, but many from widely respected contemporary scientists. Duke actively opposes polluting the white race by interbreeding, which he believes produces "dysgenic selection." He also opposes "racial intermixture," because he believes that "race suicide" could be hastened if we allow "massive immigration of an alien race" into our society and "the loss of genetic survival through racial intermixture."³³⁸ "Such views are right out of Nazi Germany and are largely repudiated by geneticists today."³³⁹

Race mixing is especially anathema to Duke, and is the reason why he is so concerned about maintaining racial segregation. Preserving the Caucasian genotype is critical, and interracial marriage, which he believes can be prevented only by separating the races, is required to prevent degeneration of the human genome. Duke even claimed that interracial marriage is genocide, and is no less terrible than what the Germans attempted to do to the Jews—

---

332 Duke, 1998, p. 109.
333 Duke, 1998, p. 109.
334 Burch and Pendell, 1945; Burch and Pendell, 1947.
335 *Pendell, 1960, pp. 20, 23.*
336 Pendell, 1960, pp. 23, 28, 116-117.
337 Pendell, 1960, p. 208.
338 Duke, 1998, p. 106.
339 Duke, 1998, p. 106.

and, he stresses, the ultimate result will be identical.[340] Preserving the Caucasian race is but a precondition for continuing its evolution to a higher level.[341]

For all of these reasons, Duke had become very concerned about what he concludes is the negative effects of all egalitarian efforts, especially integration and the push for equal schooling for the races. He concludes that the great challenge today is the "equality of the races" question—and that in order to move up the evolutionary ladder, humans have to become smarter and healthier, and cross genetic thresholds that will someday make traveling to the moon and other feats routine.[342] Duke believes that Darwinism and racism are both clearly essential to the future of Western society. Consequently, he became highly motivated to oppose all egalitarian efforts, and rather, to support both segregation and the "advancement" of Caucasians.

Duke stresses evolution as the cause of the many contrasting traits of Caucasians and Negroids. For example, when researching evolution, he compared the behavior of "Negroids" and Caucasians, pointing to the 1975 fight between Muhammad Ali and Caucasian Chuck Wepner as an example (even though it took Ali 15 rounds to win the fight by a technical knockout). He speculated that Ali had an "evolutionary advantage" in the fight, adding that "I was probably the only one in the neighborhood who thought about the evolutionary racial differences between Ali and Wepner as the replay of the fight came on TV."[343]

Those involved in racist movements often fall for the belief that, not only are "Negroids" inferior, but Jews are as well. Not surprisingly, Duke also became an anti-Semite. After applying "evolutionary biology to the development of the Jewish people"[344] he concluded that Jews are inferior for many of the same reasons used by Hitler. Small wonder that Duke became actively involved with the American Nazi party.

Duke argued that "Charles Darwin, in his study of the changing and evolving character of all life forms, demonstrated that principles of heredity combined with" natural selection developed the exceptional abilities of mankind itself. Furthermore, the subtitle of Darwin's "masterpiece, *The Origin of Species* expresses his whole idea in a nutshell; *The Preservation of Favoured Races in the Struggle for Life.*"[345] Duke took notice that Darwin's subtitle applies

---

340 Duke, 1998, pp. 108-109.
341 Duke, 1998, p. 110.
342 Duke, 1998, p. 110.
343 Duke, 1998, p. 97.
344 Duke, 1998, p. 450.
345 Duke, 1998, p. 640.

natural selection not only at the individual level, but "even more importantly, on the selection process involving species and sub-species" meaning races, as the subtitle of his "masterpiece" demonstrates.[346]

H. G. Wells' Influence on Duke

Duke's introduction to Darwinism occurred early in his life. He stated that one of the first books his father gave him to read in grade school was H. G. Wells' classic, *The Outline of History* (1922). Wells was a life-long crusader for Darwinism ever since he was introduced to the theory in college by his famous mentor, Darwin's bulldog, T. H. Huxley. *The Outline of History*, as Duke correctly notes, attempts to defend not only Darwinism, but also state-supported use of eugenics to breed superior humans.[347]

Wells believed that evolution is an essential element in the rise and fall of nations: absorbing their conquered foes leads to dysgenics, and begins the process that leads to a nation's fall because the superior victors intermarry with the inferior losers, producing an inferior progeny—as a result, the conquerors will eventually themselves be conquered. Duke took note of Wells' key Darwinian point: "great people arise having intelligence, strength, and ambition," and create a powerful society and conquer their less-fit neighbors. Soon the "process of absorbing the conquered in their nation-state" occurs and the

> traits that originally led them to victory and dominance are lost as they gradually absorb the defeated population. Invariably the process begins again, and another people come on the scene and conquer, only to once more be absorbed by those they had vanquished. . . it became obvious to me that the race factor is present in the rise and fall of every civilization. In fact, in every fallen civilization there had been a racial change from the original founding population. The only real justification for the survival of a nation is a racial one—the survival of that specific population as a distinct genetic entity, as a source for the next generation.[348]

The racist, eugenically-oriented evolutionary views of H. G. Wells convinced Duke when he was still a young man that race was *central* to evo-

---

346 Duke, 1998, pp. 450-451.
347 Duke, 1998, pp. 118-119.
348 *Wells, 1922, p. 118.*

lutionary advancement. Duke came to conclude that his crusade against the black race is a matter of the very survival of America, a nation that he repeatedly states he loves.[349] Although a disciple of Wells, Duke is actually working for much more moderate goals than his master. Wells had no qualms about admitting his solution to the world's problems—a radical eugenics program that openly involved killing inferior beings. Wells' attitude can best be summarized in his statement that, "there is only one sane and logical thing to be done with a really inferior race, and that is to exterminate it."[350] Duke came to embrace the more moderate evolutionary views of E. O. Wilson, who

> offered powerful evidence that behavior in the most elementary creatures such as ants ... to the complexities of mankind itself, had a biological basis driven by the urge to preserve the genotype. Genetic kinship turned out to be a powerful factor in evolution and behavior. In such a context, group loyalty and altruism became understandable from the evolutionary perspective in that the individual may sacrifice his life and his individual reproduction to ensure the survival to those who are genetically similar to him.[351]

As shown in this statement by Duke, the ideas of 'kin selection' and 'inclusive fitness,' along with Dawkins' notion of "selfish genes," were also critically influential to his beliefs. Genocidal or not, they are all racist views promoted by Darwinists.

Other Evolutionists Who Influenced Duke

Of the many persons whom Duke lists that influenced his racist views, most were professional Darwinists, including Julian Huxley and George Bernard Shaw.[352] He also studied the books of Henry Garrett, former Chair of the Psychology Department at Columbia University and President of the American Psychological Association (APA), and *African Genesis* by Robert Audry.[353] Duke also relied on Sir Arthur Keith's "dynamic" book, *A New Theory of Human Evolution* (1949), which stressed that not only individuals, but also groups (such as racial groups) are subjected to evolutionary pressures.

Duke even relied on Frances Galton's writings, the man who coined

---

349 Duke, 1998, pp. 118-119.
350 quoted in Trombley, 1988, p. 32.
351 Duke, 1998, p. 451.
352 Duke, 1998, p. 640.
353 Zatarain, 1990, pp. 79, 88.

the term "eugenics," and endeavored to control human reproduction to improve "the inborn qualities of a race."[354] Duke observed that Darwin wrote to Galton, openly giving complete support to Galton's eugenic views. He decided that relying on great evolutionists such as Darwin and Galton (as well as Harvard professors Wilson, Hooten, Coon and others, including "many of the leading lights of Western Civilization") lent scientific support to his ideas, empowering him to carry on his anti-black campaign with confidence and vigor.[355]

Many biological works completed by well-known scientists whom Duke had read have been reprinted by various modern racist groups. One example is University of Texas at Austin Professor Roger J. Williams' book, *Free and Unequal: the Biological Basis of Individual Liberty*, originally published in 1953 by the University of Texas Press and reprinted by Liberty Press, a racist organization. The book emphasizes that races, whether in mice, rats, horses, insects, or humans, have all developed by evolution—and that "if human beings failed to develop races they would constitute the only exception in the whole biological kingdom."[356]

Williams teaches that, although Caucasians and other races can interbreed, this does not prove equality. Furthermore, he underscores his Darwinian belief that the whole basis of evolution is variability and that some human variations are superior to others. In Williams' words, "variability is at the very basis of human life and of all life. The concept of evolution as we have it today is one in which variation is absolutely indispensable. Without genetic variability, evolution could not possibly have happened, and in line with currently accepted thought, biology itself would not exist!"[357]

This variability is what evolution selects from, in the Darwinian view—and while Williams' work is mildly racist compared to many, the racist implications are clear—which is why it was reprinted by Liberty Press. Professor Williams is clear that the writings of Darwin and his nephew, Galton, were the basis of eugenics. Williams admits that their ideas on improving the race did not have the advantage of knowing "how complicated heredity is," and they "not only flew in the face of religious teaching but were so oversimplified that they came to be regarded as unsound scientifically."[358] Nevertheless, he implies that a more sophisticated analysis of the problem may lead us

---

354 Duke, 1998, p. 640.
355 Duke, 1998, p. 640.
356 Williams, 1953, p. 210.
357 Williams, 1953, p. 56.
358 Williams, 1953, pp. 314-315.

to a practical, workable eugenics program.

The books that Duke cites as being vital in the development of his ideas all rely heavily upon Darwinism. For example, one of the most notorious racist books in the last century, Putnam's *Race and Reason: A Yankee View* (1961), published by the prestigious Public Affairs Press of Washington D.C., has laudatory introductions by numerous leading American scientists, including Ruggles Gates, Ph.D., Henry Garrett, Ph.D., D.Sc., Robert Gayre, D.Sc. and Wesley C. George, Ph.D., all eminent Darwinian scientists.

The forward by T. R. Waring states that Dr. Gates is "generally acknowledged to be one of the world's leading human geneticists."[359] Gates was a zoology professor at University of California for many years, and ended his career as an honorary research fellow of biology at Harvard. Gayre was Editor of *Mankind Quarterly*, Professor of Anthropology and head of the *Post-graduate Department of Anthropo-Geography* at the University of Sugaor in India. His many publications include a three-volume set titled *Ethnology*.

Wesley George was Professor of Anatomy at the University of North Carolina, where he was the Department Head for a decade. He was also the author of many articles on human evolution and other vertebrates. Waring states: "there can be no doubt that the endorsement of these men, taken together with the evidence of other scientists called as witnesses by the author in his text, guarantee the scientific integrity of *Race and Reason* and confirm the soundness of its premises."[360] This staunchly Darwinian book "began Duke's intellectual journey" as the most infamous living racist.[361]

Not satisfied with quoting anthropologists and Darwinists who agreed with his racist position, Putnam also attacked scientists who disagreed with him. Foremost among them was Franz Boas and his students, especially Ashley Montague and Gene Weltfish. Boas was one of the first anthropologists to openly and actively oppose the eugenics movement and the attempts to base racism on science. Putnam dismissively labeled Boas and his disciples as "the father[s] of equalitarian anthropology in America."[362] Duke is still active in racial politics. He was in the news recently for endorsing Donald Trump, then reversing his endorsement when Trump condemned extremist groups by name — including neo-Nazis and the Ku Klux Klan.[363]

---

359 Putnam, 1961, p. iv.
360 Putnam, 1961, p. v.
361 Putnam, 1961, p. 256.
362 Putnam, 1961, p. 23.
363 *Miller. 2017.*

References

Bergman, Jerry. 1993. "Evolution and the Origins of the Biological Race Theory." *CEN Tech Journal*, Vol. 7(2):155-168.

_____ and Jeffrey Tomkins. 2012. "Is the human genome nearly identical to chimpanzee?--a reassessment of the literature." *Journal of Creation.* 25(4):54–60.

Bridges, Tyler. 1994. *The Rise of David Duke.* Jackson, MS: University of Mississippi.

Brown, Harwood. 2000 (reprint edition). *Papers Read at the Meeting of Grand Dragons Knights of the Ku Klux Klan At their First Annual Meeting held at Asheville, North Carolina, July 1923.* North Stratford, NH: Ayer Company Publishers.

Browne, Janet. 2003. "The Flood, the Ark, and the Shaping of Natural History" Chapter 5 in Lindberg and Numbers.

Burch, Guy Irving and Elmer Pendell. 1945. *Population Roads to Peace or War.* Washington, D.C.: Population Reference Bureau.

_____ . 1947. *Human Breeding and Survival: Population Roads to Peace or War.* New York: Penguin Books.

DeGobineau, Arthur. 1966. *The Inequality of Human Races: The Pioneering Study of the Science of Human Races.* Los Angeles, CA: The Noontide Press.

Duke, David. 1998. *My Awakening: A Path to Racial Understanding.* Covington, LA: Free Speech Press.

Fry, Henry P. 1922. *The Modern Ku Klux Klan.* Boston, MA: Small, Maynard & Company.

Gitlin, Marty. 2009. *The Ku Klux Klan: A Guide to an American Subculture.* Santa Barbara, CA: Greenwood.

Gould, Stephen Jay. 1977. *Ontogeny and Phylogeny.* Cambridge, MA: Harvard University Press.

Keith, Sir Arthur. 1949. *A New Theory of Human Evolution.* New York: Philosophical Library.

Lee, Martin A. 2003 "Detailing David Duke." Southern Poverty Law Center. 2 pp. http://www.splcenter.org/get-informed/intelligence-report/browse-all-issues/2003/spring/detailing-david-duke

Lindberg, David and Ronald Numbers. 2003. *When Science and Christianity Meet*. Chicago, IL: University of Chicago Press.

Maginnis, John. 1992. *Cross to Bear*. Baton Rouge, LA: Darkhorse Press.

Martinez, J. Michael. 2007. *Carpetbaggers, Cavalry, and the Ku Klux Klan: Exposing the Invisible Empire During Reconstruction*. New York: Rowman & Littlefield.

Miller, Haley. 2017. Ex-KKK Leader David Duke Has Meltdown After Trump Condemns White Supremacists In Charlottesville. *Huffpost*, August 14. https://www.huffingtonpost.com/entry/david-duke-trump-charlottesville_us_5991d6bae4b08a2472764798

Morris, Charles. 1888. *The Aryan Race: its Origins and its Achievements*. Chicago: S. C. Griggs and Company.

Nelson, G. Blair. 2003. "Men Before Adam! American Debates Over the Unity and Antiquity of Humanity." Chapter 7, pp. 161-181, in Lindberg and Numbers.

Newton, Michael. 2007. *The Ku Klux Klan: History, Organization, Language, Influence and Activities of America's Most Notorious Secret Society*. Jefferson, NC: McFarland & Company, Inc., Publishers.

_____. 2010. *The Ku Klux Klan in Mississippi: A History*. Jefferson, NC: McFarland.

Pendell, Elmer (editor). 1942. *Society Under Analysis, an Introduction to Sociology*. Lancaster, PA: Cattell.

_____. 1951. *Population on the Loose*. New York: Wilfred Funk.

_____. 1960. *The Next Civilization*. Dallas, TX: Royal Publishing Company.

_____. 1967. *Sex Versus Civilization*. Los Angeles, CA: Noontide Press.

_____. 1977. *Why Civilizations Self-Destruct*. Cape Canaveral, FL: Howard Allen Enterprises.

Putnam, Carleton. 1961. *Race and Reason: A Yankee View*. Washington, D.C.: Public Affairs Press.

Rose, Douglas D. (editor). 1992. *The Emergence of David Duke and the Politics of Race*. Chapel Hill: University of North Carolina Press.

Topinard, Paul. 1894. *Anthropology.* London: Chapman & Hall.

Tomkins, Jeffrey and Jerry Bergman. 2012. "Genomic monkey business—estimates of nearly identical human-chimp DNA similarity re-evaluated using omitted data." *Journal of Creation.* 25(4):94-100.

Trombley, Stephen. 1988. *The Right to Reproduce: A History of Coercive Sterilization.* London: Weidenfeld and Nicholson.

Turner, John. 1982. *The Ku Klux Klan: A History of Racism and Violence.* Montgomery, AL: The Southern Poverty Law Center.

Wells, H. G. 1922. *The Outline of History.* New York: Collier.

White, Adam. 1966. *The Negro ... Animal or Human?* Alexandria, VA: Adam White.

Williams, Roger J. 1953. *Free & Unequal: The Biological Basis of Individual Liberty.* Indianapolis, IN: Liberty Press.

Winchell, Alexander. 1978. *Proof of Negro Inferiority.* Metairie, LA: Sons of Liberty.

Wise, Tim. 2003. *Great White Hoax: Responding to David Duke and the Politics of White Nationalism.* Seattle: Northwest Coalition for Human Dignity.

Zatarain, Michael. 1990. *David Duke: Evolution of a Klansman.* Gretna, LA: Pelican Publishing.

Chapter 18

Albert Einstein, Ph.D.

From Christian to Agnostic

"Science without religion is lame, religion without science is blind."

Albert Einstein is yet another example of someone who "had respect for the religious values enshrined within Judaic and Christian traditions,"[364] but became an agnostic due to his study of Darwinism. He was born into a Jewish family that was not particularly observant, choosing not to follow traditional Jewish dietary laws or attend religious services. His parents sent Albert to a Catholic public primary school at age six, but he also received instruction in Judaism from a distant relative. When Einstein moved on to the Luitpold Gymnasium, he received two hours of religious instruction each week that the school offered its Jewish pupils. Einstein studied such topics as the Ten Commandments and biblical history. He was also very active in proselytizing for his faith, even writing scriptures on buildings with chalk.

Albert Einstein is often regarded as one of the most eminent scientists who ever lived. His name alone is a symbol of genius, as in the phrase "Who do you think you are, Einstein?" Furthermore, "modern physics bears his impact more than that of [any] other physicist[s] [re-check quote]."[365] As a Jew who experienced much persecution in Nazi Germany, he was also a great humanitarian, deeply concerned about the lives and suffering of the people around him.[366] A man of many talents, he even developed a skill level on the violin so high that he was qualified to perform with leading orchestras. His popularity is also indicated by the fact that over 412 biographies have been written about Einstein and his work.[367]

Einstein wrote in his autobiographical notes that he left his once devout childhood faith as a result of reading popular scientific books about evolution and other science topics. When he started reading secular science books, especially those that dealt in some way with Darwinism, Einstein found himself agreeing with the Darwinian worldview. As we have seen in so

---

364 Randerson, James, 2008.
365 Jammer, 1999, p. 3.
366 Seeger, 1982.
367 Jammer, 1999, p. 3.

many other cases, he decided that if Darwin was right, much of the

> Bible could not be true. The consequence was a positively fanatic freethinking coupled with the impression that youth is intentionally being deceived by the state through lies; it was a crushing impression. Suspicion against every kind of authority grew out of this experience, a skeptical attitude toward the convictions that were alive in any specific social environment - an attitude which has never again left me, even though later on ... it lost some of its original poignancy."[368]

Einstein believed in a "God" but not the personal God of Christianity or Judaism, rather, in his words, he believed in Spinoza's God, a god that reveals himself in the orderly harmony of the universe and its laws, and not in a living God who is concerned with the fates and actions of humans. He once proclaimed, "I cannot imagine a God who rewards and punishes the objects of his creation, whose purposes are modeled after our own — a God, in short, who is but a reflection of human frailty. Neither can I believe that the individual survives the death of his body, although feeble souls harbor such thoughts through fear or ridiculous egotisms."[369] He was clear to denounce religious beliefs which he regarded as "childish superstitions."[370]

> The word god is for me nothing more than the expression and product of human weaknesses, the Bible a collection of honourable, but still primitive legends which are nevertheless pretty childish. No interpretation no matter how subtle can (for me) change this.[371]

In 1936 Einstein received a letter from a young girl who wrote to him on behalf of her Sunday school class.

> *Dear Dr. Einstein, we have brought up the question: 'do scientists pray?' In our Sunday school class. It began by asking whether we could believe in both science and religion. We are writing to scientists and other important men, to try and have our own question answered.*

---

368 Einstein, 1970, p. 5. Also, Albert Einstein, quoted in his *The New York Times* obituary, April 19, 1955; from George Seldes, ed., *The Great Thoughts*, New York: Ballantine Books, 1996, p. 134.
369 Albert Einstein, quoted in George Seldes, ed., 1996, p. 134.
370 Randerson, James, 2008.
371 Randerson, 2008

> *We will feel greatly honored if you will answer our question: do scientists pray, and what do they pray for?*
>
> *We are in the 6th grade, Miss Ellis's class.*
>
> *Respectfully yours,*
>
> *Phyllis*

Einstein replied a few days later:

> *I will attempt to reply to your question as simply as I can. Here is my answer:*
>
> *Scientists believe that every occurrence, including the Affairs of human beings, is due to the laws of nature. Therefore a scientist cannot be inclined to believe that the course of events can be influenced by prayer, that is, by a supernaturally manifested wish.*
>
> *However, we must concede that our actual knowledge of these forces is imperfect, so that in the end the belief in the existence of a final, ultimate Spirit rests on a kind of faith. Such belief remains widespread even with the current achievements in science.*
>
> *But also, everyone who is seriously involved in the pursuit of science becomes convinced that some spirit is Manifest in the laws of the universe, one that is vastly superior to that of man. In this way the pursuit of science leads to a religious feeling of a special sort, which is surely quite different from the religiosity of someone more naive.*
>
> *With cordial greetings,*
>
> *A. Einstein* [372]

Einstein elsewhere declared "I am not an atheist,"[373] and preferred to consider

---

[372] Hendricks, 2017.
[373] Isaacson, 2008, p. 390.

himself an agnostic, or a "religious non-believer."[374]

References

Calaprice, Alice. 2000. *The Ultimate Quotable Einstein.* Princeton, New Jersey: Princeton University Press.

Hendricks, Scotty. 2017. "Did Einstein Pray? What the Great Genius Thought About God." *Big Think.* http://bigthink.com/articles/did-einstein-pray-what-the-great-genius-thought-about-god

Isaacson, Walter. 2008. *Einstein: His Life and Universe.* New York: Simon and Shuster.

Jammer, Max. 1999. *Einstein and Religion.* Princeton, New Jersey: Princeton University Press.

Randerson, James. "Childish superstition: Einstein's letter makes view of religion relatively clear." *The Guardian,* 13 May 2008.

Seeger, Raymond J. 1982. "Einstein, Cosmotheist." *Journal of the American Scientific Affiliation*, pp. 42-44, March.

Seldes, George, ed., 1996. *The Great Thoughts*, New York: Ballantine Books.

Video

Albert Einstein – The History Channel (1:29:52)
https://www.youtube.com/watch?v=R_yk45m4E3M

---

374 Calaprice, 2000.

Chapter 19

Rachel Held Evans

From YEC to Christian Evolutionist

Eighty years after the Scopes Monkey Trial spectacle that brought national attention to her hometown, Rachel Held Evans began to doubt her own faith. Evans grew up with the many advantages of Christianity, including a Christian home, private schooling, a theologian dad and her school's 'Best Christian Attitude Award' for four consecutive years.

Although she grew up in a culture she claims was obsessed with apologetics, Evans asked a lot of questions which indicates she had very inadequate knowledge in the area of apologetics. She writes that she "used to be a fundamentalist in the sense that I thought salvation means having the right opinions about God and that fighting the good fight of faith requires defending those opinions at all costs."[375]

After she was exposed to Darwinism, she began to feel that, in order for her faith to survive, she must adapt (actually conform) like biological evolution alleges. Her own spiritual journey from certainty, through doubt, to a "modern" faith in Darwinism, illustrates the ongoing struggle about Darwinism and the church that has challenged the Christian community in recent years.

Her cynical book, *Faith Unraveled*, is the story of the loss of both her creationism belief and her jettisoning of the basic Christian doctrines. Evans achieved an enormous amount of positive publicity for her book, including a favorable article in *USA Today*. Her book garnered a whopping 174 reviews on Amazon, almost all very laudatory, and only 1 percent gave it a 1 or 2 rating.

Evans' harsh judgments of the Christian world in which she grew up and her cynicism pervade the book's narrative. She claimed she was taught at Bryan College that evolution fails to account for the

> complexity inherent in biological organisms, to produce a sufficient fossil record of transitional species, and to explain the many ambiguities in biological classification. Therefore, it should not be taught as fact in public schools. Most important, the theory of evolution is dangerous because it undermines the authority of the

---

375 Evans, 2014, p. 17.

Bible and threatens the foundation of Christianity.[376]

All of this, she claims, she has now rejected, including all that she learned

> from Dr. Kurt Wise, one of the leading young-earth creationists in the country and a favorite professor among Bryan College students... Armed with a degree in paleontology from Harvard University, Dr. Wise had studied under Stephen J. Gould, a renowned evolutionist and science writer. Dr. Wise said that his goal was to formulate a model of earth history consistent with both Scripture and the scientific data.[377]

She added that

> While not everyone on campus supported young-earth creationism, the overriding principle behind the school's educational approach was that the Bible serves as our most reliable textbook, that it provides an infallible foundation on which to build the academic disciplines. We learned that everything from science to history, economics, art, psychology, politics, and literature can be studied from a "biblical worldview."[378]

She would soon also reject all of these beliefs, and her lack of knowledge often showed up in her book when she tried to explain why. An example included naively asking, "What if Galileo had simply accepted church-instituted cosmology paradigms," a statement that showed her major lack of knowledge about this central historical event.[379] She obtained much of the information that caused her to accept evolution from the mass media. One of her favorite TV series of all time was BBC's very biased series on evolution titled Planet Earth. [380] She loved the show because narrator David Attenborough

> shows how magnificently living organisms can adapt to their environments. From the extra-thick eyelashes of the wild Bactrian camel, to the dense white fur of the Arctic hare, to the sticky yellow toes of the gliding leaf frog, each animal has its unique way of

---

376 Evans, 2014, p. 70.
377 Evans, 2014, p. 70.
378 Evans, 2014, p. 71.
379 Evans, 2014, p. 219.
380 Evans, 2014, p. 210.

> thriving in its habitat, be it a dusty desert, a snowy tundra, or the tops of trees.[381]

This evidence is less proof of Darwinism than it is of Intelligent Design, first, because they are only examples of microevolution that creationists already accept, and second, because mutations have no foresight. Only a pre-designed mechanism could make existing designs function in a changing environment.

Other evidence Evans cited in favor of evolution that caused her to reject creationism, included cave angelfish. Marine angelfish are known for their attractive colorful fins in aquariums, but because cave angelfish

> lost the pigment in their skin, they look more like ghostly, winged snakes than fish. Microscopic hooks on their fins allow them to cling to cave walls like bats. Positioned just right, they can feed on bacteria rushing down the waterfalls. Their eye sockets are empty, their bodies elongated and slimy.[382]

Her arguments for evolution include the claim that scientists

> believe that a group of marine angelfish must have migrated to the caves millions of years ago to escape predators or to adjust to climate change. Like many cave dwellers, over the years they evolved blindness and improved other sensory functions because in their environment eyesight was no longer useful. Cave angelfish live exclusively in a few remote caves in Thailand, and the Planet Earth crew went to all kinds of trouble to capture footage of these little survivors in their natural habitat.[383]

Loss of function of cave animals (for example, loss of sight) is not a problem for creation, but the gain of function is a major problem for evolution.

For her conversion to evolutionism, Evans relied strongly on social media —hardly a reliable source of solid scientific information. In one blog, she "encountered all kinds of people, young and old, who no longer considered themselves Christians because of false fundamentals" such as one person who "took a biology class and was convinced that the theory of evolution was

---

381 Evans, 2014, pp. 210-211.
382 Evans, 2014, p. 211, emphasis added.
383 Evans, 2014, p. 211. emphasis added.

sound," and another person who opined that she was "tired of fighting the culture wars."[384] She added, "I looked into the science behind evolution,"[385] asking, "why are we so afraid to confront the mountain of scientific evidence in support of evolution?"[386]

She admitted that her experience at Bryan College "was everything a college experience should be. I made lifelong friends, learned how to think critically, and became well-versed in Christian apologetics."[387] As is blatantly obvious from her writings, it appears that she was *not* very well versed in Christian apologetics. She claims her teachers at Bryan College put the students into the predicament of learning that science "contradicted the evolution paradigm, but [we] engaged in some mental gymnastics of our own, trying to explain how it's possible to see the light from distant stars."[388] She must not have learned that this is a puzzle for all astronomers, not just creationists.

Evans admitted that, "My story isn't pretty" and it isn't finished yet, but she is "telling it because it's the best evidence I've got in support of … evolution." While asserting that Christianity must adapt to Darwinism,[389] she never even hints in her book where the mountains of evidence are in favor of evolution. *Faith Unraveled* is a tragic example of what happens when one is poorly grounded in the evidence against Darwinism. The lure of evolutionary "fake science" that leads to a rejection of creation opens up a void of confusion, contradiction, and conflicted thoughts.

Reference

Evans, Rachel Held. 2014. Faith Unraveled: How a Girl Who Knew All the Answers Learned to Ask Questions. Grand Rapids. Zondervan.

Video

Asking Better Questions (21:31)
https://www.youtube.com/watch?v=9Ob-0eOSkqw
Talks about how she experimented with the Biblical blueprint for how to be a godly woman.

---

384 Evans, 2014, p. 208.
385 Evans, 2014, p. 97.
386 Evans, 2014, p. 96.
387 Evans, 2014, p. 78.
388 Evans, 2014, p. 79.
389 Evans, 2014, p. 212.

Chapter 20
Martin Gardner, Ph.D.
From Christian Fundamentalist to Agnostic

One of the most prolific popular science writers of the last century and an early promoter of the modern skeptical movement was Martin Gardner (1914-2010). Gardner was reared a "Fundamentalist Christian" who later became an apologist for evolution, and an anti-creationist. He called Darwin Doubters of all types "cranks who are paranoid and consider themselves geniuses, but nobody else who is normal does."[390] Thus, in his view, about 80 percent of the American population are cranks. His wife was Jewish, only because she has a Jewish background, and Martin had a Seventh-day Adventist background, until he discovered Darwin. As a result, both ended up outside of any particular church tradition, and neither attends a church or synagogue. While still living, Gardner wrote that he now calls himself

> a fideist, who believes something on the basis of emotional reasons rather than intellectual reasons...As a fideist I don't think there are any arguments that prove the existence of God or the immortality of the soul. Even more than that, I agree ... that the atheists have the better arguments. So it is a case of quixotic emotional belief that is really against the evidence and against the odds. The classic essay in defense of fideism is William James' *The Will to Believe*. James' argument, in essence, is that if you have strong emotional reasons for a metaphysical belief, and it is not strongly contradicted by science or logical reasons, then you have a right to make a leap of faith if it provides sufficient satisfaction.[391]

During Gardner's religious period of his life, he writes that a counselor at a camp that he attended for several summers, George Getgood,

> introduced me to literature published by the Moody Bible Institute, in Chicago. I was strongly moved by a book of sermons by Dwight L. Moody, America's most famous evangelist before

---
390 Gardner. 1957, pp. 11-12.
391 Shermer, 1997.

the days of Billy Sunday. I recall being especially impressed by Moody's sermon on the blood of Jesus. It was the only part of Jesus' body, Moody said, that was left on Earth. Christ's blood, he argued, runs like a scarlet thread through both Testaments. It also drips through the pages of my religious novel, The Flight of Peter Fromm.[392]

In a chapter in his autobiography titled, "I Lose My Faith," Gardner covers how and why he lost his Christian faith.[393] After that time, he wrote "with acute embarrassment," that he was once involved in "Seventh-day Adventism, a fundamentalist sect still flourishing around the world."[394] In his autobiography,[395] he details his journey from believer to deist.

A central reason why he left both the SDA Church and Christianity was, as he writes, "When I first entered the University of Chicago I was in the grip of a crude Protestant fundamentalism," and not long after this he wrote that he became a deist.[396] He explained that it

> was a course in Geology 101 that widened the cracks in my crumbling faith. In high school, during my flirtation with Adventism, I found in Tulsa's library a big textbook called The New Geology by Adventist George McCready Price. Price persuaded me, to the sorrow of my geologist father, that the earth was about ten thousand years old and fossils were stone forms of life that perished in the great flood. A minimum knowledge of geology from my 101 course convinced me that Price was an amiable crank.

This is an example of the name-calling that is very common among those evolutionists who oppose creationism. In fact, Price was a professor at several colleges including Loma Linda University, and wrote 30 best-selling books, mostly on geology. Some of his conclusions may be wrong, but he was not by any stretch of the imagination stupid or "an amiable crank." As to the claim, "He had no training in geology,"[397] Price in fact taught geology at the college level for decades, and also wrote extensively about it. Gardner concluded that in college he found himself

---

392 Gardner, 1957, p. 53.
393 Gardner, 1957, p. 53.
394 Gardner, 1957, p. 54.
395 Gardner, 2013.
396 Gardner, 2013, p. 53.
397 Gardner, 2013, p. 57.

faced with the following dilemma. If evolution is true, as I came to believe, then Genesis is not. And if Genesis is false, how could I trust the accuracy of other biblical events. If Eve wasn't fabricated from Adam's rib—and what a droll insult to women that myth is!—then perhaps Jesus never raised Lazarus from the dead after his body had started to decompose.[398]

He soon moved from theistic evolution to rejecting Christianity altogether, writing that during his first year at the University of Chicago he

> was one of the founders of the Chicago Christian Fellowship, a small band of fundamentalists very much out of place at a secular university ... I continued to attend its weekly gatherings even after my faith had started to waver. At one meeting I spoke about the truth of evolution, and how one could accept it without giving up faith in God and Christianity.[399]

He now realizes that the goal of accepting two contrary ideas, such as Darwinian evolution and Christianity, does not work. One of his professors, the great American author Thornton Wilder,

> was a theist, the son of missionaries to China. I was astounded to hear Wilder say on what he called a "digression day"—a day on which he responded to student questions—that he had not yet made up his mind on whether Jesus was uniquely divine or just a great religious teacher.[400]

Also of interest is the fact that

> Wilder's best seller novel, *The Bridge of San Luis Rey*, is about a bridge in South America that collapses, killing a number of pedestrians. Father Juniper researches the lives of those who perished in an effort to determine whether their deaths could somehow have been planned by God. He reaches no final conclusion.[401]

---

398 Gardner, 2013, p. 58.
399 Gardner, 2013, p. 58.
400 Gardner, 2013, p. 59.
401 Gardner, 2013, p. 59.

Gardner's family also influenced his beliefs. He writes: "I don't know if my mother believed in evolution, but I do know my father did. I found among his papers a clipping from a Tulsa paper about a speech he gave at a Rotary Club luncheon in 1923."[402] The article's

> headline is "Story of Creation Told in the Bible Is a Contribution to Mythology, He Believes." Although Dad praises the Bible as a great "rule and guide of life," and says Jesus was in some sense divine, he makes clear that the earth is hundreds of millions of years old, and the Genesis account of creation must be replaced by a recognition of the slow evolution of all forms of life. God, he said, is "universal love and truth and goodness."[403]

In this case, as is true of so many others, the age of the Earth was a major factor that influenced his progression from creationist to evolutionist. Another was the problem of evil and suffering. When Martin Gardner died on May 20, 2010 at age 94, the magazine Skeptical Inquirer devoted an entire issue to him that included laudatory articles by no less than 18 leading scientists and academics.[404] He is widely considered to be on the list of the top ten most outstanding and influential skeptics of the twentieth century.

References

Martin Gardner. 1957. *Fads & Fallacies in the Name of Science*. New York: Dover.

_____. 2013. Undiluted Hocus-Pocus: *The Autobiography of Martin Gardner*. Princeton, NJ: Princeton University Press.

Shermer, Michael. 1997. "The Annotated Gardner." *Skeptic Magazine* Vol. 5, No. 2. https://www.skeptic.com/eskeptic/10-05-26/#tribute

---

402 Gardner, 2013, p. 101.
403 Gardner, 2013, p. 101.
404 September-October, 2010, 34(3):28-42.

Chapter 21
Karl W. Giberson, Ph.D.
From YEC to Veneer Christian Agnostic

Karl Giberson, formerly Professor of Physics at Eastern Nazarene College, is now a professor at Stonehill College in Easton, Massachusetts, a Catholic college. He was also the former Vice President of the theistic evolutionary BioLogos Foundation. He writes that, in 2009, he visited the Creation Museum in Petersburg, Kentucky, because he

> was in the area to speak at Xavier University, where a class taught by a Jesuit theologian was using my book *Saving Darwin: How to Be a Christian and Believe in Evolution*, a deeply personal account of my own—and America's—struggle to understand Darwin's controversial theory of our origins.[405]

He writes his Creation Museum visit was bittersweet for the reason that

> the beautiful story of creation told in the museum's lovingly constructed dioramas had once been at the heart of my religious beliefs, growing up in a Baptist parsonage among the potato fields of New Brunswick, Canada. As a youth, uncharacteristically obsessed with theology and biblical studies, I had dreamed of one day working with organizations like Answers in Genesis that had built the Creation Museum. My goal when I enrolled at an evangelical liberal arts college in 1975 was to train in science and join the creationist cause.[406]

This ambition was soon derailed when he was taught in college that molecules-to-humankind evolution was proven to be unequivocally true by scientific fact, and that "creationism was scientifically indefensible," a view he now accepts somewhat uncritically. As he explored the Creation Museum's

> displays of Adam and Eve, their temptation and sin, and their expulsion from Eden, I felt sadness. I wondered about my younger

---
405 Giberson, 2015, p. 1.
406 Giberson, 2015, pp. 1-2.

self, the debater forever defending creationism, convinced that the only [true] origins story ... was the one in Genesis, that Adam and Eve were as historical as Abraham Lincoln and Winston Churchill, that modern scientific theories of origins were without foundation and inspired by Satan to lead people away from God's truth. That younger man now seemed like someone else, a stranger I once knew.[407]

He became very active working hard to convince others that Darwinism is true and all forms of creation, including Intelligent Design, are false. For example, under the headline, "Why Evangelicals Are Fooled into Accepting [the] Pseudoscience," of Adam and Eve and other myths, Giberson claims that one strategy employed by evangelicals in their "crusade against evolution" is "to undermine the entire scientific enterprise." He was obviously not very familiar with the empirical research that Darwin Doubters use to defend their worldview. Giberson also incorrectly claims that creationists argue that if

> science is a deeply flawed, ideologically driven, philosophically suspect enterprise, then why should anyone care if almost every scientist supports the theory of evolution? If the scientific community is just a bunch of self-serving ideologues with Ivy League appointments, then we can ignore anything it says that we don't like.[408]

This view, he concludes, is foolish because he was a true believer of the view that science has demonstrated evolution from cells to humans to be a fully documented fact. Even though Giberson was "a vexatious critic of biblical creation... [he] feigned surprise that so many of his students ended up leaving the church altogether."[409] This perfectly illustrates one of the key points I make in this book that evolution leads students down a slippery slope towards atheism once they embrace theistic evolution. As Giberson admitted, most evangelical colleges now teach evolution as fact

> albeit quietly, carefully, and often tentatively, although there are exceptions. ... Those of us teaching evolution at evangelical colleges are made to feel as if we have this subversive secret we must whisper quietly in our students' ears: 'Hey, did you know that

---

407 Giberson, 2015. pp 1-2.
408 Giberson. 2011.
409 Sarfati, 2015

Adam and Eve were not the first humans and never even existed? And that you can still be a Christian and believe that?"[410]

How he does this is explained by John West[411] who writes that, after debating Giberson, he observed that Giberson was at one time "a serious participant in discussions of issues related to science and faith. A co-founder of the BioLogos Foundation with Francis Collins and Darrel Falk ... of late it's getting hard to take him seriously." An example he provided is, during public lectures, Giberson actually used

> a fake photo purporting to show a human baby born with a tail. Giberson used the photo to illustrate his claim that human babies are occasionally "born with perfectly formed, even functional tails," which is supposed to provide evidence of humans' shared evolutionary ancestry with lower animals. It turned out that not only was Giberson's scientific claim bogus -- so was his photo. The picture was a Photoshopped concoction he downloaded from a humor website.[412]

West adds that

> Giberson's use of the photo in public lectures may say something about his lack of skepticism when it comes to claims made in the name of evolution ... after the photo was exposed, he did eventually apologize -- albeit with very poor grace and a stream of self-justifications. He even accused my colleague David Klinghoffer, who helped expose the fake photo, of "willful lies," which was itself a patent falsehood. The whole episode was a sorry example of Giberson's scorched-earth and fact-free approach to those he disagrees with.[413]

The result was Giberson

> posted an over-the-top bromide against Discovery Institute at *The Huffington Post* under the overwrought title ... "Discovery Institute Still Undermining Science." The article makes Giberson's

---
410 Giberson, 2014. In this article, Giberson laments poll results indicating an uptick in public acceptance of creation and the corresponding negative response to evolution.
411 West. 2015.
412 West. 2015.
413 West. 2015.

> statements about humans born with tails seem a model of probity. Inveighing against Discovery Institute's "teach the controversy" education policy, Giberson asserts that the "controversies" the Institute wants taught in science classes include a 10,000-year-old earth, Noah's Ark, and Adam and Eve.[414]

West adds that Giberson has provided no documentation for these claims "because they are absolutely false," adding that he [West] has "been involved with Discovery Institute before the Institute even had its program on Intelligent Design, and we've never advocated teaching the things he says in science class... The Discovery Institute doesn't even favor teaching about Intelligent Design in K-12 classes." West concluded that we do not "support banning the teaching of evolution, despite Giberson's additional false claim that our 'real agenda' is 'to get evolution out of the public schools.'"

> On the contrary, we think science students should learn more about evolutionary theory, not less. That includes the best evidence for modern evolutionary theory, but it also includes scientific disputes over key evolutionary claims already being aired in mainstream peer-reviewed science journals. These include disputes over the creative power of the mutation-selection mechanism: How much can natural selection acting on random mutations actually accomplish?[415]

West correctly adds that "If Giberson disagrees with the criticisms of Darwinian theory raised by scientists in the Intelligent Design community, he should take the time to respond to those criticisms rather than spread falsehoods." For example

> Giberson could respond to the scientific claims made in Stephen Meyer's book Darwin's Doubt. Giberson may think Meyer raises no serious scientific claims, but a number of other scientists disagree, ranging from Harvard geneticist George Church to paleontologist Mark McMenamin, co-author of *The Emergence of Animals* (Columbia University Press). McMenamin even called Meyer's book "a game changer for the study of evolution and points us in the right direction as we seek a new theory for the origin of animals." Neither Church nor McMenamin is affiliated with the

---

414 West. 2015.
415 West, 2015.

Intelligent Design movement, so it took a great deal of courage for them to recommend Meyer's book.

The fact is

> rather than respond to the real issues raised by those affiliated with Discovery Institute, Giberson simply knocks down straw men. That's the sort of intellectual evasion I've grown to expect from the hard-core partisans of modern evolutionary theory. I had thought that Giberson was better than that.

West concluded that

> Giberson's fact-free attacks may play well with the Darwin Amen chorus at The Huffington Post and The Daily Beast, but surely he is capable of something more. Perhaps Giberson has convinced himself he lives in an alternate universe -- a universe where a Discovery Institute exists that actually promotes the things he claims. But for those of us who live in the real universe, I wish he'd rejoin us. He might be an interesting discussion partner.

Giberson's response titled, "My Debate with an 'Intelligent Design' Theorist"[416] was in the *Daily Beast*. Notice that he uses quotes around the words 'Intelligent Design.' His main assertion in this article is that ID does not have a "theory." In fact, it has several, including the principles that complex specified information always requires an intelligent cause, that irreducible complexity is unattainable by chance mechanisms, and that the finely-tuned constants of physics are best explained by purposeful design.

---

416 Giberson, 2014.

References

Giberson, Karl. 2015. *Saving the Original Sinner: How Christians Have Used the Bible's First Man to Oppress, Inspire, and Make Sense of the World*. Boston: Beacon Press.

———. 2014. "2013 Was a Terrible Year for Evolution." *thedailybeast*, January 2. https://www.thedailybeast.com/2013-was-a-terrible-year-for-evolution

———. 2014. "My Debate With An 'Intelligent Design' Theorist." *thedailybeast*, April 21. https://www.thedailybeast.com/my-debate-with-an-intelligent-design-theorist

———. 2011. "Why Evangelicals are Fooled Into Accepting Pseudoscience." *Huffpost*, November 23. http://www.huffingtonpost.com/karl-giberson-phd/evangelicals-and-science_b_975821.html

Video: *Should Christians Embrace Darwin?* Debate between Stephen C. Meyer and Giberson (2:14:23) https://www.youtube.com/watch?v=9sXsRa8vDpw

Sarfati, Jonathan. 2015. "Evolution Makes Atheists Out of People – Mark 2! Ex-Biologos leader Karl Giberson admits futility of his compromise." *Creation*, December 3. https://creation.com/evolution-makes-atheists-out-of-people-mark-2

West, John. 2015. "The Sad Decline of Karl Giberson." *Evolution News*, November 9. http://www.evolutionnews.org/2015/11/the_sad_decline100731.html

Video: *Controversies in Science* (2012) with Michael Shermer (56:46) https://www.youtube.com/watch?v=4-Krxa1q00w

Chapter 22

Philip Dean Gingerich, Ph.D.

From Amish to Evolutionist

Philip Dean Gingerich, Ph.D. (1946-) is Professor Emeritus of Geology, Biology, and Anthropology at the University of Michigan, Ann Arbor. He was Director of the Museum of Paleontology at the University of Michigan from 1981 to 2010. Dr. Gingerich was reared by a family of Amish Mennonites in eastern Iowa. His grandfather was a farmer and a lay preacher. He received an A.B. from Princeton University in 1968, a M.Phil. from Yale University in 1972, and a Ph.D., also from Yale, in 1974, all in the field of geology.

He is most well-known for his extensive field research attempting to prove whale evolution. In an interview with *National Geographic*, he said that "I grew up in a conservative church in the Midwest and was not taught anything about evolution. The subject was clearly skirted. That helps me understand the people who are skeptical about it because I come from that tradition myself."[417] In the same interview, he noted that becoming an evolutionist has, for him, been "a spiritual experience."

His search for the origin of mammals led to field work in South Asia, starting in the 1970s, which has yielded fossil evidence of several new land mammals in Pakistan. One, which he named Pakicetus (Pakistan whale), he considers a putative link between a whale ancestor and modern whales. His ongoing field work in the Tethyan sedimentary sequence of central Nepal has yielded other fossils that he claims are intermediate forms supporting the origin and early evolution of whales. He now spends his life trying to convince others that evolution is true, and creationism is false. Unfortunately for evolutionists, his research has totally failed to make a clear, convincing case for the evolution of whales from some terrestrial small dog-like mammal. Some of the problems include artistic reconstructions showing flippers without fossil evidence for them, which Gingerich later admitted were not flippers. In my view, he has promulgated speculative scenarios as science because of an active imagination and an agenda to promote evolution.

---

417 Quammen, 2004.

Reference

Quammen, David. 2004. The Evidence for Evolution is Overwhelming. *National Geographic*. 206(5): p. 31, November.

Videos

Dismantling Whale Evolution (9:47)
https://www.youtube.com/watch?v=5eltxBT670M

Out of the Blue (7:54) https://www.youtube.com/watch?v=AVxZUaPsc0Q
Discussing the relationships between whales, goats, and sheep.

How Fast is Evolution? (1:01:07)
https://www.youtube.com/watch?v=AVxZUaPsc0Q

Chapter 23

Stephen Godfrey, Ph.D.

A Creationist Becomes an Aggressive Anti-Creationist

Dr. Stephen Godfrey is Paleontologist and Curator at the Calvert Marine Museum located in Solomons, Maryland. Twenty-five years ago, the study of fossils

> set Godfrey on an anguished path. Raised in a fundamentalist Christian family in Quebec, Canada, embracing a 6000-year-old Earth where Noah's flood laid down every fossil, Godfrey began probing the underpinnings of creationism in graduate school. The inconsistencies he found led ..., over many years, to a staunch acceptance of evolution. With this shift came rejection from his religious community, estrangement from his parents, and, perhaps most difficult of all, a crisis of faith that endures.[418]

Professor Couzin quotes Godfrey claiming that, "emotions bind together young-Earth creationists," which are "members of a movement making inroads from Kenya to Kentucky."[419] Godfrey's faith was a solid part of his upbringing. For example, after dinner, his father, a Sunday school teacher, would pull "out the Bible. 'We would go systematically through two readings of books,' ... and devote time to prayer. The family attended church twice on Sundays, in the morning and in the evening, and one parent ... dropped in on a Bible study class midweek."[420]

He added that, for helping religious students accept evolution, Godfrey considers it unfortunate that "Scientists and educators have responded mainly by boosting biology's place in the classroom and building rational arguments for evolution."[421] He feels that reason alone is rarely enough to sway creationists because, for them, abandoning a creation worldview

> carries enormous emotional risks, including a loss of identity and community and an agonizing, if illusionary, choice: science or

---

418 Couzin, 2008. p. 1034.
419 Couzin, 2008, p. 1034.
420 Couzin, 2008, pp. 1034-1035.
421 Couzin, 2008, p. 1034.

faith. People ... tend not to advertise their painful transition from creationist to evolutionist, certainly not to scientific peers. When doubts about creationism begin to nag, they have no one to turn to: not Christians in their community, who espouse a literal reading of the Bible and equate rejecting creationism with rejecting God, and not scientists, who often dismiss creationists as ignorant or lunatic.[422]

From a young age, Godfrey had a very strong interest in biology, which he expressed by touring natural history museums and collecting objects like pinecones, rocks, minerals and especially skeletons in particular. Because his "parents saw no conflict between their son's love of biology and their Christian beliefs, they encouraged his interests. "I guess they figured that the young-Earth creationist position was strong enough, was robust enough, that he would believe in young-Earth creationism and he would be a biologist, and that would be fine," Godfrey remarked.[423] When Godfrey was age 48, he recalled that, when a child, he grew up when

> young-Earth creationism took hold in North America in the early 1960s. Its leaders argued that during the previous 150 years, Bible-believing Christians had gone too far in accommodating science in their interpretation of scripture and pushed for a literal reading of the Bible ... Fossils, for example, are the remains of plants and animals left out of Noah's ark. The description of Adam and Eve in Genesis suggests that humans had never been subject to evolution. Using calculations drawn from genealogy, young-Earth creationists consider the planet to be 6000 to 10,000 years old. (Geologists say it is about 4.5 billion years old.)[424]

At this time, Godfrey had

> subscribed wholeheartedly to these views, vividly recalls his earliest encounter with evolution. In the first grade ... a student teacher said that apes were the ancestors of people. "I remember having this visceral reaction ... and saying, 'No, that can't be.'" Around the dinner table that night, his family discussed the experience, concluding that the teacher must have been mistaken. "It couldn't

---

422 Couzin, 2008, p. 1034.
423 Couzin, 2008, pp. 1034-1035.
424 Couzin, 2008, p. 1035.

be true because apes aren't evolving into humans today; they're apes," Godfrey remembers.[425]

Godfrey suffered an identity crisis when he began college because he still was "convinced that scientists were engaged in a vast conspiracy to promote evolution."[426] He majored in biology at Bishop's University in Sherbrooke, Quebec, and lived at home. He claimed that, in one sense, his science studies had little effect on his faith, writing that

> "You can learn facts, and you can do really well on exams and not believe" what you're learning ... But then his classes also raised niggling questions that biblical literalism could not easily answer. For example, ... A literal reading of Genesis indicates that no animals perished before Adam and Eve ate the fateful apple—in other words, that there were no carnivores preying on other animals.[427]

Godfrey's doubts grew when he learned in his biology classes that predators are perfectly designed

> to kill: cats with stereoscopic vision, enlarged canines, and claws; spiders that weave webs as traps; and sharks that replace serrated teeth throughout their life. "They're not eating seaweed," says Godfrey, who puzzled over how these animals had emerged if God hadn't intended them to prey on others. "That was the first thing at university that really started to disturb me," he says.[428]

This issue has been dealt with numerous times in creationist writings, information which he should at least have acknowledged, and, ideally, attempted to deal with. In his last year of college, Godfrey presented a talk

> on the origin of flight, arguing that *Archaeopteryx*, the earliest known bird, could not have evolved from the dinosaurs. Although impressed by similarities between *Archaeopteryx*'s anatomy and that of dinosaurs, he pushed this to the back of his mind. By this time, Godfrey was ... determined to find out for himself whether the claims of biologists and paleontologists were true. He enrolled

---

425 Couzin, 2008, p. 1035.
426 Couzin, 2008, p. 1035.
427 Couzin, 2008, p. 1035.
428 Couzin, 2008, p. 1035.

in graduate school in paleontology at McGill University and was taken in by Robert Carroll. Carroll had heard that Godfrey was a creationist but didn't give it much thought, he says now. In Carroll's lab, Godfrey prepared and described fossils of an ancient amphibian called *Greererpeton*. The fossils "could have come from the moon," says Carroll. Analyzing them out of context had little impact on Godfrey's views.[429]

Eventually, Godfrey's worldview "came crashing down" when he was invited to join an expedition to rural Kansas involving University of Toronto paleontologist Robert Reisz and some students

> digging for pelycosaurs, 300-million-year-old animals that display some features of mammals that evolved later. By day, quarrying through thin layers of rock, "we started to come across footprints of terrestrial animals," said Godfrey. "You can't imagine a global flood and animals finding ground to make footprints on. ...That, more than anything, any other experience in my life, really shook me to the core." Godfrey agonized about where these footprints might have come from. Some creationists argue for floating mats of vegetation during the flood, but Godfrey found that unconvincing.[430]

Arguments besides the floating mats theory exist, and Godfrey should have researched them instead of assuming no other good arguments could be examined and considered. His professor on the field trip commented that Godfrey "was one of the brightest students" that he had ever had and if this was true he certainly could have done the research as others have done to answer his question. Professor Reisz "at the time knew that Godfrey was a devout Christian but had no idea of the crisis triggered by his fieldwork. The ease with which he learned, and the ease with which he accumulated new ideas all spoke to a superior intelligence."[431] Nor did Godfrey explain what triggered his crisis on a field expedition. Godfrey did not yet fully accept evolution until he moved

> to Drumheller, Alberta, dubbed the "dinosaur capital of the world" because of its diversity of fossils. Godfrey often drove

---

429 Couzin, 2008, p. 1035.
430 Couzin, 2008, p. 1035.
431 Couzin, 2008, p. 1035.

southeast to Dinosaur Provincial Park, passing through a landscape of sediments laid atop one another: deposits from freshwater and terrestrial environments in one, marine organisms and mollusks in another, and a third that mimicked the first, a mix of fossils from fresh water and land. "These animals were living here in this same place, but they couldn't have all been there at the same time," he says, a fact that was irreconcilable with flood geology. It was then that "the rest of the young-Earth creationist ideas kind of exploded."[432]

According to Couzin, Godfrey expressed "bitterness, anger, and disappointment about having been deceived for so many years." This case is actually more the failure of Godfrey to use his "superior intelligence" to seek out answers to his questions, which many creationists have had, some of which have been written about in detail. One response of Godfrey to his anger was to seek out creationists in order to confront them with his new conclusions. One example he gives was when still

> in graduate school, he and his devout Christian wife, mother-in-law, and mother attended a weekend symposium at a Bible school in New York state, where Godfrey says he angrily stood up at the end of a talk and argued passionately with the speaker. It was there, and in conversations during holiday meals, that Godfrey's parents realized that he had changed. Deeply unhappy, they worried whether their son could endorse an old Earth and remain a Christian. Their message was, "It's all or nothing," says Christopher Smith, Godfrey's brother-in-law and a pastor at the University Baptist Church in East Lansing. Michigan. "I do remember a discussion one year at Christmas; the tone quickly turned angry," Smith says. Godfrey's father eventually asked that he stop mentioning evolution, as the topic was too upsetting to the family, who believe that their afterlife depends on embracing creationism.[433]

When attempting to articulate exactly where his religious conclusions now stood

> Godfrey's eyes fill with tears. "It's been so long, a lifelong struggle, to sort out," ... He has flirted with atheism but found it too depressing. Several years ago, he stopped attending church for a year

---

432 Couzin, 2008, pp. 1035-1036.
433 Couzin, 2008, p. 1036.

before returning. He believes in God today, he says, but tomorrow may be different. Complicating matters are the people he most loves and their stance on creationism. Godfrey and his wife met as teenagers in a church youth group. They and their five children have always attended an evangelical, young-Earth creationist church.[434]

One other experience was when

Godfrey seethed through 12 weeks of a DVD presentation on creationism.[435] During an early session, he raised objections in front of a church youth group that included his 15-year-old daughter. The group was not brought back for later showings. "I was really torn," he says, "because I would have loved to have been given the opportunity to say, 'Okay, I'm now going to do a presentation on the other side.' But they don't want to hear it. It's too threatening and it's too upsetting." Like many creationists-turned-evolutionists, Godfrey is conflicted about how, and how forcefully, to press his case.[436]

In 2005, Godfrey and his brother-in-law, Christopher Smith, published the book titled *Paradigms on Pilgrimage*,[437] that makes the case for evolution and also describes their transition from their YEC worldview to Darwinism. Godfrey writes his father prayed that their book

would not be published, and Godfrey did not send his parents a copy. He thought his book would change minds among creationists but isn't sure it has. "I haven't" read it, says his younger sister Esther Godfrey, of Sherbrooke. ... "it's a very odd way of viewing the Bible, if you can choose which parts you believe literally and not literally." Esther Godfrey is not sure what turned her brother away from a young Earth, as they've never discussed it. "I know he saw something at some point, maybe a fossil, and thought the Earth has to be old," she says. Just as he longs for biblical literalists to be more receptive to evolution, Godfrey also wishes that biologists would join the discussion. He was incensed 5 years ago when, participating in an evolution-creationism debate at Bishop's

---

434 Couzin, 2008, p. 1036.
435 He should have given the title and specifically what he objected to here.
436 Couzin, 2008, p. 1036.
437 Godfrey, 2005.

University, where he once argued against the fossil record, no one from the biology department attended.[438]

Couzin then adds that some former creationists believe that

> changing minds is not worth the heartache it brings. Godfrey no longer considers evolution worth mentioning to his parents ... "You can live your life just fine and not know squat about evolution," he says. When it comes to his children, Godfrey's not sure what they believe nor how firmly to steer them. Certainly, he says, they are exposed to creationist teachings. Of all his children it's his youngest, 4-year-old Victoria, who shows the strongest penchant for science. Wandering the beaches near her home, she often asks to bring home bones she finds, just as her father did years ago. Will her view of the world make room for evolution? Godfrey watches and waits and wonders.[439]

References

Couzin, Jannijer. 2008. "Crossing the Divide" *Science* 319(5866):1034-1036. *February 22.*

Godfrey, Stephen J. and Christopher R. Smith. 2005. *Paradigms on Pilgrimage: Creationism, Paleontology and Biblical Interpretation.* Toronto: Clements Publishing.

---

438 Couzin, 2008, p. 1036.
439 Couzin, 2008, p. 1036.

Chapter 24

Hiram Bentley Glass, Ph.D.

Creationist to Leading Evolutionary Scientist

Hiram Bentley Glass (January 17, 1906 – January 16, 2005) was a leading American geneticist and a popular science columnist, writing and editing several books and over 400 scientific articles. He also regularly wrote a science column in *The Baltimore Evening Sun*. Glass had a distinguished career, not only as a writer, but also as a scientific policy maker and theorizer on provocative and often prescient ideas about major policy issues, such as genetics and nuclear war, that still concern us today.

Born in China to missionary parents, he attended Decatur Baptist College, then the Baptist institution Baylor University, and lastly the University of Texas, where he earned a Ph.D. under the mentorship of the eminent, but controversial evolutionary geneticist, Hermann Joseph Muller, who is famous for his studies on mutations.[440] His *New York Times* obituary noted that, despite "his Baptist upbringing, he developed an early and deep interest in evolution as a result of a fascination with insects."[441]

Glass rejected creation due to, he claims, the overwhelming scientific evidence for evolution, a topic in which he had "a deep interest." He never mentions what this evidence is in the sources that I have read about him. As a leading evolutionist, he has published widely in the peer-reviewed scientific literature. Obviously, his Baptist education did very little to help him critically analyze the many major problems with Darwinism even though both Decatur and Baylor were more conservative than they are today.

Dr. Glass was also at one time or another President of the American Society of Human Genetics, the American Society of Naturalists, the American Institute of Biological Sciences, Chairman of the Atomic Energy Commission's Advisory Committee for Biology and Medicine, advisor at Cold Spring Harbor Laboratory, and he was a member of the National Research Council's Space Science Board (SSB).

---

440 Erk, 2005.
441 Douglas, 2005.

References

Erk, Frank C. 2005. "Remembering Bentley Glass (1906–2005)." *The Quarterly Review of Biology.* 80(2):165-173. June.

Douglas, Martin. 2005. "H. Bentley Glass, Provocative Science Theorist, Dies at 98." *New York Times.* January 20.

Chapter 25

Richard Goldschmidt, Ph.D.

From Creationist to Leading Evolutionist

Another example of a well-known scientist who converted to atheism as a result of studying Darwinism is the late Richard Goldschmidt, Professor Emeritus of Biology at the University of California, Berkeley. He wrote that when he read books by German evolutionist Ernst Haeckel, it seemed to him that "Evolution was key to everything and could replace all the beliefs and creeds"[442] of religion, and that no supernatural creation event had ever occurred in all of history, except

> evolution and the wonderful law of recapitulation which demonstrated the fact of evolution to the most stubborn believer in creation. I was so fascinated and shaken up that I had to communicate to others my new knowledge ... I remember vividly a scene during a school picnic when I stood surrounded by a group of schoolboys to whom I expounded the gospel of Darwinism as Haeckel saw it ... my zeal, which was Haeckel's zeal, was that of a missionary.[443]

His wholehearted conversion to Darwinian evolution seems to have solidified when he gave his first public presentation

> on evolution (of course a dramatic digest of Haeckel) before a boys' club.... There is no doubt that hundreds of thousands, young and old, inside and outside Germany, were impressed in the same way [as I was], and that thus Haeckel[444] became one of the most beloved and most hated men of his time.[445]

Shortly thereafter, Goldschmidt lost his childhood Christian faith and, he stated, started hanging around with "bad company"—he joined a

---

442 Goldschmidt. 1956, p. 35.
443 Goldschmidt. 1956, p. 35.
444 It's worth pointing out that Haeckel was a heavy promotor of illustrations that he deliberately faked depicting side by side embryonic development of different species that were incredibly similar, leading to the acceptance of evolution among millions of unsuspecting people for many decades.
445 Goldschmidt. 1956. pp. 35–36.

gang, lied, stole, drank, and "flunked a grade."⁴⁴⁶ The public dishonor to his parents and others that resulted from behaving, in Goldschmidt's words, "abominably," shocked him into changing his lifestyle.⁴⁴⁷

Later, Goldschmidt realized that Haeckel's case was far less persuasive than he had first assumed. By the time Goldschmidt became a world-famous zoologist, he recognized Haeckel was "a fanatic, bigoted zealot" who forged his most famous evidence for Darwinism (in Goldschmidt's words, Haeckel "improved upon nature"). Haeckel ended up in much trouble over his various forgeries, especially his embryo series that made it appear as if all humans pass through a fish, reptile, and other stages of evolution in the womb.⁴⁴⁸ By then, though, Goldschmidt was hooked: "Darwin, Haeckel, Huxley, Spencer became my heroes."⁴⁴⁹

Ironically, when Goldschmidt came to believe from his research that there were "some major errors in present day evolutionary theory" he was "violently attacked" by his Darwinist critics.⁴⁵⁰ He added that, as a result of his criticism of Neo-Darwinian gradualism, the

> Neo-Darwinians reacted savagely. This time [the Darwinists thought] I was not only crazy but almost a criminal. There were, of course, exceptions, like the deep thinker Sewall Wright, who criticized my work objectively and has recently moved much nearer to my views.⁴⁵¹

His heresy was realizing that evolution could not occur by a gradual means as Darwin taught, but must have happened in jumps due to macro-mutations. He called the product of these mutations "hopeful monsters," because mutations typically produce monstrous defects, if the creature survives at all. He imagined that a few of the mutations might have produced major evolutionary leaps—something Darwin had explicitly denied. This theory has not only been disproven today, but is an embarrassment to evolutionists. Yet the problems that Goldschmidt documented, such as the many gaps in the fossil record, still remain true to this day. In fact, those gaps are much more problematic today than when Goldschmidt was alive.

---

446 Goldschmidt. 1960, pp. 27–28.
447 Goldschmidt. 1960, p. 28.
448 Goldschmidt. 1956, pp. 32-35.
449 Goldschmidt. 1960, p. 29.
450 Goldschmid. 1960, p. 307.
451 Goldschmidt. 1960, p. 324.

References

Richard Goldschmidt. 1940. *The Material Basis of Evolution*. Yale university Press. From the Silliman lectures he presented at Yale University in 1939. http://www.evolocus.com/Textbooks/Goldschmidt1960.pdf

_____. 1960. *In and Out of the Ivory Tower: The autobiography of Richard B. Goldschmidt*. Seattle, WA: University of Washington Press.

Richard Goldschmidt. 1956. *Portraits from Memory: Recollections of a Zoologist*. Seattle: University of Washington Press.

Chapter 26

Ernst Haeckel, Ph.D.

Creationist to Darwin's Nazi Disciple

Outside of Darwin, one of the most famous Darwinists in Europe, especially the Germanic–speaking world, was Professor Ernst Haeckel. This German Darwinist was influential in leading Germany down the road to Nazism and the Holocaust.[452] Haeckel was a "conservative, religious youth" who "metamorphosed in adulthood into a scientific leader of radical inclinations— in politics as well as science."[453] Haeckel states that he was reared "by pious parents who belonged to the Free Evangelical church." During the first twenty years of his life he was "a convinced and zealous adherent of" Christianity.

His transformation from Christian to atheist began during his college classes in the Department of Natural Science and Medicine (1852-1857). Later, through his travels, he "gradually reached, through heavy soul conflicts, the conviction that the mystic faith-teachings of the Christian religion were completely irreconcilable with the certain results of scientific experience."[454] As a result, he stated that it was utterly impossible for him to reconcile "the Christian beliefs about 'creation', etc., with the important facts of evolution."[455] The influence of Darwinism and his study of the world, specifically his cosmology, changed him so drastically that he moved away from his *dualistic* and idealistic worldview to a purely *monistic* philosophy, meaning the material world is all that exists. No human souls, gods, angels, or spirits exist. A decisive influence in this direction for Haeckel included the writings of

> Darwin. The fundamental lines (Grundzuge) of a strictly monistic ... philosophy ... especially from the teachings of evolution, I ... outlined in my "General Morphology of Organisms"— my introductory work, in 1866; and later, in a more popular form, largely from that work in my "Natural History of Creation," in 1868.[456]

Haeckel wrote that the course of his life "in its third decade...thor-

---

452 Bergman, 2012.
453 Forrest and Gross, 2004, pp. 103–104.
454 McCabe and Wakeman, 1911, pp. 27–28.
455 McCabe and Wakeman, 1911, pp. 27–28.
456 McCabe and Wakeman, 1911, pp. 28–29.

oughly convinced me that the Christian religion, as far as the ethical and practical affairs and conduct of life were concerned" was "unreliable and unsatisfactory in every point." He also recalled that for over "twenty years I had *inwardly*, from pure conviction, absolved myself from the faith-teachings of Christianity" and consequently,

> it would have been only natural to have given proper expression to this conviction outwardly by withdrawal from the Evangelical church. But this last step I left untaken out of regard to my family and some dear friends to whom I thereby would have brought heavy sorrow and injury.[457]

The writings of Haeckel were a powerful stimulus for the rise of Nazi Germany, as documented by Bergman and others.[458] His influence was also important in converting thousands of persons to Darwinism especially through his many books, which sold many thousands of copies.[459]

References.

Bergman, Jerry. 2012. *Hitler and the Nazi Darwinian Worldview: How the Nazi Eugenic Crusade for a Superior Race Caused the Greatest Holocaust in World History*. 2012. Kitchener, Ontario, Canada: Joshua Press.

Forrest, Barbara and Paul Gross. 2004. *Creationism's Trojan Horse: The Wedge of Intelligent Design*. New York: Oxford University Press.

McCabe, Joseph and Thaddeus Burr Wakeman. 1911. *The Answer of Ernst Haeckel to the Falsehoods of the Jesuits*. New York: The Truth Seeker Company.

Spiro, Jonathan. 2009. *Defending the Master Race*. Burlington, VT: University of Vermont Press. p. 123.

---

[457] McCabe and Wakeman, 1911, pp. 29–30.
[458] Bergman, 2012.
[459] Spiro. 2009, p. 123.

Chapter 27

Stephen Hawking, Ph.D.

Journey from Christian to Dogmatic Atheist

Professor Stephen Hawking, (1942-2018) an international celebrity, has sold millions of books, and drew huge crowds wherever he spoke. Cited by *Time Magazine* as the intellectual heir to Einstein, he was arguably surpassed only by Darwin and Einstein as best-known science figures by the public. The first American edition of his best seller, *A Brief History of Time*, had a press run of ten thousand copies—typical mainline press runs are from five hundred to two thousand copies.[460] His later popular book did almost as well.[461]

As a Cambridge University professor, he occupied the Lucasian Professor of Physics Chair, the same Chair that Isaac Newton had filled almost three centuries earlier. Hawking's research was on the mathematical physics of black holes. In 1978, he received the most prestigious honor in physics, the Albert Einstein Award, equivalent to the Nobel Prize.[462] In his many best-selling books he explained modern cosmology theory with elegance, fluidity, precision, and accuracy.

Hawking is not only famous as a physicist, but also as one who has overcome many obstacles due to his severely disabling neuromuscular disease, Amyotrophic Lateral Sclerosis (ALS), commonly called Lou Gehrig's Disease after the famous baseball player who died from it. He was diagnosed while still in college, and lived far beyond anyone's expectations, including the leading medical experts.[463]

Courtship and Marriage

His wife, Jane Wilde, married Stephen knowing then that he had an incurable disease. They first met at a New Year's Eve party when she was still in high school, and "there is little doubt that" meeting Jane Wilde was a major turning point in "Hawking's life. The two of them began to see a lot more of one another and a strong relationship developed. It was finding Jane Wilde

---
460 Hawking, 1988.
461 Hawking, 2001.
462 White and Gribbin, 1992, p. 187.
463 DeLange, 2012, p. 27.

that enabled him to break out of his depression and regenerate some belief in his life and work."[464] For Hawking

> his engagement to Jane was probably the most important thing that ever happened to him: it changed his life, gave him something to live for and made him determined to live. Without the help that Jane gave him, he would almost certainly not have been able to carry on, or had the will to do so.[465]

Believing that his life would be very short, they hoped to fit as much love and fulfillment as possible into what they thought would be, at the most, only a few years together. They married fairly young, and soon had three children. Stephen outlived all expectations, and they were together for over a quarter of a century.

Jane turned out to be an astoundingly supportive caregiver, effectively dealing with Stephen's progressive physical decline and the heavier demands that his illness placed on her. She managed the household, reared the children, and hauled him around for years before a serious respiratory incident forced them to hire full-time professional nurses. She also battled both the British health care system and Cambridge University for wheelchair access. Her motivation, as reviewed by Professor Schaefer, was because "Jane Hawking is a Christian," and

> "Without my faith in God, I wouldn't have been able to live in this situation (namely, the deteriorating health of her husband, with no obvious income but that of a Cambridge don to live on). I would not have been able to marry Stephen in the first place because I wouldn't have had the optimism to carry me through, and I wouldn't have been able to carry on with [out this optimism]". [466]

One problem in their marriage was a result of the domestic friction that one would expect when a family member is seriously handicapped. A more serious problem, though, was their religious conflicts. Their marriage also provided much insight into the age-old conflicts between science and religion.

---

464 White and Gribbin, 1992, pp. 63-64, 70.
465 White and Gribbin, 1992, pp. 63-64, 70.
466 Schaefer, 2003, pp. 58-59.

Jane's Theism vs. Stephen's Atheism

A factor central to their relationship conflicts—and eventual divorce—was religious differences. As a youth, Hawking was reared a Christian, but as his interest in science grew, he concluded that "there was no crumb of comfort to be found there [in religion]" when he was given two years to live.[467] In Jane's words, "Stephen had no hesitation in declaring himself an atheist despite the strong Methodist background" of his childhood and family.[468]

Jane remarked, "as a cosmologist examining the laws which governed the universe, he could not allow his calculations to be muddled by a confessed belief in the existence in a Creator God."[469] She accepted the anthropic principle theory for God's existence, which she calls one of the most important cosmological findings of the twentieth century.[470] Her husband's social network only reinforced his atheism. Although he apparently had no problem invoking the anthropic

> principle to explain the enigmatic "fine-tunings" of the universe, many physicists considered the principle the equivalent of scientific voodoo. David Gross, a colleague of Jim Hartle's at the University of California, Santa Barbara, called it "dangerous" because it has been used by some to bolster the notion that science can provide evidence for the existence of a God who purposefully set up the universe in such a way as to allow for the eventual evolution of human beings. "It smells of religion" [he said].[471]

Hawking later even rejected the general anthropic principle.

Jane recognized that the strong anthropic principle has a "close philosophical affinity to the medieval cosmos" where humans were at the center of creation.[472] It placed humans in a special place in the universe, and, for the "medieval populist, this special position was a strong statement of the unique relationship between human beings and their Creator."[473] The main intent of the early philosophers was to reconcile the "existence of God with the rigors of the laws of science, towards unifying the image of the Creator with the

---

467 White and Gribbin, 1992, p. 62.
468 Hawking, 2004, p. 46.
469 Hawking, 2004, p. 46.
470 Hawking, 2004, p. 153.
471 Larsen, 2007, pp. 103-104.
472 Hawking, 2004, pp. 153
473 Hawking, 2004, pp. 153-154.

scientific complexity of His Creation" but

> their intellectual heirs, some 800 years later, seemed intent on distancing science as far as possible from religion and on excluding God from any role in Creation. The suggestion of the presence of a Creator God was an awkward obstacle for an atheistic scientist whose aim was to reduce the origins of the universe to a unified package of scientific laws, expressed in equations and symbols. To the uninitiated, these equations and symbols were far more difficult to comprehend than the notion of God as the prime mover, the motivating force behind Creation.[474]

She added that, as a direct result of modern cosmologists focus on mathematics, the concept of a personal God became irrelevant for many scientists because, in their mind, their calculations diminished "any possible scope for a Creator," and

> they could not envisage any other place or role for God in the physical universe. Concepts which could not be quantified in mathematical terms as a theoretical reflection of physical realities, whether or not the actual existence of those physical realities was proven, were meaningless.[475]

Although her husband talked about knowing "the mind of God" in his bestselling book, he was not referring to a personal god as taught by Jews, Christians and Muslims, but rather a scientific understanding of "the theory of everything," meaning the unification of the four forces, the strong and weak nuclear forces, and gravity and electromagnetism, which are the basis of holding everything in the universe together.[476] His god was the natural world, yet Hawking once admitted that he is still attempting "to understand how the Universe works, why it is the way it is and why it exists at all. I think there is a reasonable chance that we may succeed in the first two aims, but I am not so optimistic about finding why the Universe exists."[477]

---

474 Hawking, 2004, pp. 154-155.
475 Hawking, 2004, pp. 154-155.
476 Dowe, 2005, pp. 142-143.
477 quoted in White and Gribbin, 1992, p. 291.

Atheism Nihilism

Jane's major concern was her perception, based on discussions with her husband and the leading physicists of the world, that a result of the goal of their beliefs about science included atheism. This goal would eventually include reducing "human reactions in all their complexities, emotional and psychological... to scientific formulae because, in effect, these reactions were no more than the microscopic chemical interactions of molecules."[478] The result was that "in the face of such dogmatically rational arguments, there was no point in raising questions of spirituality and religious faith, of the soul and of a God who was prepared to suffer for the sake of humanity—questions which ran completely counter to the selfish reality of genetic theory."[479]

As a result of her interactions with her husband and his colleagues, Jane realized that, "at the end of the twentieth century, religion finds its revolutionary truths threatened by scientific theory and discovery, and retreats into a defensive corner, while scientists go into the attack insisting that rational argument is the only valid criterion for an understanding of the workings of the universe."[480] She witnessed that the complexity of the cosmologist's calculations and the admiration of their discoveries have caused some people

> to fall into the trap of believing that science has become a substitute for religion and that, as its great high priests, they can claim to have all the answers to all the questions. However, because of their reluctance to admit spiritual and philosophical values, some of them do not appear to be aware of the nature of some of the questions.[481]

Jane, who herself is an intelligent woman with a Ph.D., was especially disheartened with attempts by evolutionists to extrapolate the rules of animal behavior to human behavior, particularly in the field of evolutionary psychology. After noting that evolutionary psychologists ascribe altruism solely to natural selection, she responded that

> scientists still cannot satisfactorily explain why some human beings are prepared to give their lives for others. The complexity of

---
478 Hawking, 2004, p. 156.
479 Hawking, 2004, p. 156.
480 Hawking, 2004, p. 200.
481 Hawking, 2004, p. 200.

such anomaly lies far outside the scope of their purely mechanical grasp. Nor can they explain why so much human activity operates at a subliminal level. The spiritual sophistication of musical, artistic, politic, and scientific creativity far exceeds that of any primitive function programmed into the brain as a basic survival mechanism.[482]

Although evolutionists offer rescue devices for this problem, Jane noticed that they "acknowledge that they are still very far from reaching" the goal of answering "why" such behavior exists. From her experience with evolutionists through her husband's friends, she remarked that they

> arrogantly even aspire to become gods themselves by denying the rest of us our freedom of choice and disputing our right to ask the question "Why?" in relation to the origins of the universe and the origins of life. They claim that the question is ... as inappropriate as it would be to ask why Mt. Everest is there. They dismiss the suggestion that the question "why" is the prerogative of theologians and philosophers rather than scientists because, they say, theologians are engaged in the "study of fantasy": belief in God can be attributed to "a shortage in the oxygen supply to the brain."[483]

The imaginations of her husband and his friends reduced the entire universe

> to a handful of material components. They complain with a weary disdain of the stupidity of the human race, that human beings are always asking "Why?" Perhaps they should be asking themselves why this is so. Might it not be that our minds have been programmed to ask "why?" And if this is the case they might then ask who programmed the human computer. The "Why" question is the one which, above all, theologians should be addressing.[484]

She concludes by opining that, since the modes of thought by scientists are dictated by purely

> materialistic criteria, physicists cannot claim to answer the ques-

---

482 Hawking, 2004, p. 200.
483 Hawking, 2004, p. 201.
484 Hawking, 2004, p. 201.

tions of why the universe exists and why we, human beings, are here to observe it, any more than molecular biologists can satisfactorily explain why, if our actions are determined by the workings of a selfish genetic coding, we sometimes listen to the voice of conscience and behave with altruism, compassion and generosity.[485]

Ironically, most all of Hawking's work is not empirical science, but largely speculation based on thought experiments. White and Gribbin note that Hawking is not eligible for the Nobel Prize because one of their rules is

> a candidate may be considered for a prize only if a discovery can be supported by verifiable experimental or observational evidence. Hawking's work is, of course, unproved. Although the mathematics of his theories is considered beautiful and elegant, science is still unable even to prove the existence of black holes, let alone verify Hawking Radiation or any of his other theoretical proposals.[486]

The Marriage Deteriorates

As he aged, Stephen's atheism became increasingly dogmatic, unreasonable and aggressive. Jane notes that, as a result, although in the early days their arguments on religion "were playful and fairly light-hearted," in later years Stephen's arguments increasingly

> became more personal, divisive and hurtful. It was then apparent that the damaging schism between religion and science had insidiously extended its reach into our very lives: Stephen would adamantly assert the blunt positivist stance which I found too depressing and too limiting to my view of the world because I fervently needed to believe that there was more to life than the bald facts of the laws of physics and the day-to-day struggle for survival. Compromise was anathema to Stephen, however, because it admitted an unacceptable degree of uncertainty when he dealt only with the certainties of mathematics.[487]

In the latter days of their marriage her "attempts to discuss the profound matters of science and religion with Stephen were met with an enig-

---
485 Hawking, 2004, p. 200.
486 White and Gribbin, 1992, p. 188.
487 Hawking, 2004, p. 201.

matic smile."⁴⁸⁸ Stephen usually "grinned at the mention of religious faith and belief, though on one occasion he actually made the startling concession that, like religion, his own science of the universe" also required a leap of faith.⁴⁸⁹

Jane agreed with scientist-theologian Cecil Gibbon's conclusion that "scientific research required just as broad a leap of faith in choosing a working hypothesis as did religious belief."⁴⁹⁰ Although leaps of faith in science ideally "had to be tested against observation," the problem is that a scientist must "rely on an intuitive sense that his choice was right or he might be wasting years in pointless research with an end result that was definitively wrong."⁴⁹¹

When Stephen was openly asked if he believed in God, the answer was always the same. "No," he did not and "there was no room for God in his universe."⁴⁹² During a visit to Jerusalem, Stephen gave his usual atheistic answers, a response that struck Jane as especially ironic, causing her to quip that her

> life with Stephen had been built on faith—faith in his courage and genius, faith in our joint efforts and ultimately religious faith—and yet here we were in the very cradle of the world's three great religions, preaching some sort of ill-defined atheism founded on impersonal scientific values with little reference to human experience.⁴⁹³

She regretfully admitted that Stephen's blunt denial of "all that I believed in was bitter indeed."⁴⁹⁴ In fact the "couple had lived with religious disunity for most of their married life," this fact alone was an insufficient reason to separate.⁴⁹⁵ The problem was

> Hawking's early agnosticism had become more overtly atheistic, and with his no-boundary theory he had effectively dispensed with the notion of God altogether. Yet, ironically, Jane's deeply held religious convictions had been one of the strengths which had enabled her to cope so well with the burden imposed by Stephen's

---

488 Hawking, 2004, p. 465.
489 Hawking, 2004, p. 465.
490 Hawking, 2004, p. 465.
491 *Hawking, 2004, p. 465.*
492 Hawking, 2004, p. 494.
493 Hawking, 2004, pp. 494-495.
494 Hawking, 2004, p. 495.
495 White and Gribbin, 1992, pp. 285-286.

increasing disability.[496]

Jane was also chagrined at the insensitivity of the press to matters of religious faith—they often treated it as something that, if one possesses it, it should be kept well-hidden—religious people should stay in the closet.[497]

The Galileo Irony

Ironically, Stephen's hero was Galileo—"a devout Catholic."[498] Stephen launched a personal campaign for Galileo's reinstatement by the Catholic Church, which was eventually successful. He saw this success as a "victory for the rational advance of science over the highbound antiquated forces of religion rather than as a reconciliation of science with religion."[499] The intransigence of Stephen on religion is in dramatic contrast to the many changes he made in his and others theories and ideas—for example, his theory that "contrary to all previously held theories on black holes, a black hole could radiate energy."[500]

As Stephen became more famous and increasingly renowned in his field, he associated more and more with eminent scientists, people which Jane did not find very appealing. The contrast between her old friends and the world's leading scientists who became their friends was enormous. Their old friends were able to talk intelligently about many things and showed a "human interest in people and situations."

In contrast, their new friends were as a whole "a dry, obsessive bunch of buffoons," little concerned with people, but rather very preoccupied with their personal scientific reputations. She added that "they were much more aggressively competitive than the relaxed, friendly relativists with whom we had associated in the past."[501] Their old friends' dedication to science verged on the dilettante in comparison with the 'driving fanaticism' of their new friends.[502]

Jane stressed the fact that "nature was powerless to influence intellectual beings who were governed by rational thought, [but] who could not recognize reality when it stood, bared before them, pleading for help. They

---

496 White and Gribbin, 1992, pp. 285-286.
497 Hawking, 2004, p. 525.
498 Hawking, 2004, pp. 200-201.
499 Hawking, 2004, p. 202.
500 Hawking, 2004, p. 236.
501 Hawking, 2004, p. 296.
502 Hawking, 2004, pp. 295-296.

appeared to jump to conclusions, which distorted the truth to make it fit their preconceptions."[503]

Jane's Solace was in Her Faith

Christianity permeated Jane's world, a world in which her husband and most of his scientist friends wanted no part. Jane eventually left Stephen's world, partly because of the antagonism of Stephen and his atheistic friends. She found that most famous scientists, her husband among them, were dogmatic atheists, unwilling to even reason on the evidence for design in the Universe. Stephen's atheistic scientist friends saw someone such as Jane who believed in God as an ignoramus who inhabited a world that they wanted no part of.

Jane did not want to live in Stephen's universe, one that she increasingly saw as not only unreal, but one that blocked out reality.[504] She summarized her feelings about much of the research where her husband was at the forefront, as "theorizing on abstruse suppositions about imaginary particulars traveling in imaginary time in a looking-glass universe which did not exist except in the mind of the theorists." She described this as "the demon goddess of physics."[505]

During Hawking's meeting with the Pope, the Pope allegedly said that scientists "could study the evolution of the universe," but "should not ask what happened at the moment of Creation at the Big Bang and certainly not before it because that was God's preserve."[506] Jane stated she was not impressed with this attitude because she believed that "instead of embracing the modern scientific quest for truth to its ultimate objective in glorying in the even deeper layers the mystery thus revealed, the Vatican still viewed cosmological science as a contentious issue, a threat to religious stability, which had to be contained."[507]

Jane feels that what is dangerous is the misinterpretation of, and the use to which these discoveries are put—especially, she added, by those who have an axe to grind against theism such as many of the eminent scientists she personally knew.

---

503 Hawking, 2004, p. 312.
504 Hawking, 2004, p. 389.
505 Hawking, 2004, p. 372.
506 Hawking, 2004, p. 391.
507 Hawking, 2004, p. 391.

Stephen Becomes a God

As Stephen's fame grew, more and more in the press and the public viewed him as godlike,[508] Jane responded, "I found myself telling him that he was not god. The truth was that supercilious enigmatic ... smile which Stephen wore whenever the subjects of religious faith and scientific research came up was driving me to my wit's end. It seemed that Stephen had little respect for me as a person and no respect at all for my beliefs and opinions."[509] One of her strongly held opinions was that "reason and science alone could not furnish all of the answers to the imponderable mysteries of human existence."[510] In response,

> Jane actively fought back against the unrealistic spin the press was attempting to put on their lives, and began to be (sometimes brutally) honest about their struggles in interviews. She even commented to reporters that her role in Stephen's life had become "telling him that he was not God."[511]

Hawking could easily see himself as god-like because after the death of Carl Sagan, Isaac Asimov and Stephen Jay Gould—he became one of the most well-known scientists in the world. As a measure of his popularity, his book, *A Brief History of Time*, sold 20 million copies in 20 languages, an unheard-of record in science publishing.[512]

Yet the "simple and fairly obvious" truth about him was "most unpalatable to those people who had come to believe in Stephen's immortality and infallibility."[513] The fact is, in the minds of many people, Stephen's scientific theories became "the basis for a new religion."[514] Nonetheless, Jane concluded that her religion was a "personal relationship with God and through it,...I found the germinating seeds of an incipient peace and a wholeness which I had not known for a very long time."[515] Their daughter, Lucy Hawking, said her father "will do what he wants to do at any cost to anybody else."[516] For

---

508 Larsen, 2007, p. 8.
509 Hawking, 2004, p. 536.
510 Hawking, 2004, p. 536.
511 Larsen, 2007, p. 113.
512 Schaefer, 2003, p. 59.
513 Hawking, 2004, p. 537.
514 Hawking, 2004, p. 537.
515 Hawking, 2004, p. 572.
516 quoted in White and Gribbin, 1992, p. 290.

example, he has faced serious charges of ethics violations in his personal life.[517]

As his condition deteriorated, Jane increasingly felt like a nurse taking care of a man with a body like a Holocaust victim and the needs of a child. A critical stabilizing factor in Jane's life was her church. She often talked about her minister's sermons, and how they helped her to cope with the difficulties of dealing with an invalid husband who required twenty-four-hour-a-day care. He needed to be bathed, have his teeth brushed, his hair combed, and his bodily functions taken care of like a six-month-old baby, yet he attracted worldwide notoriety wherever he went—and they traveled often, which was also a struggle.

A concern she had was "although I derived comfort from my return to the Church, it also imposed imponderable questions in my mind." One was, "what was God really asking of me? How great a sacrifice was required of me" to care for my husband.[518] She says she is now under no obligation to "promote the greater glory of Stephen Hawking."[519]

Modern Medicine to the Rescue

Although Stephen's state of health was often extremely precarious, modern medicine and twenty-four-hour nursing care—he carried his own mini-hospital with him everywhere—allowed Stephen to pursue a "hedonistic way of life, compensating ever more tenaciously for his disability, ever more assured of his own invincibility, mocking the untimely death whose grasp he had evaded."[520] What sustained Jane was trusting "in God through darkness, pain and fear."[521] When she tried to help Stephen understand the solace she obtained from her faith, and especially the scriptures, Stephen "was insulted by any mention of compassion; he equated it with piety and religious sentimentality"—something for which he had nothing but contempt.[522]

Jane had friendships with numerous well-known scientists, and many were close personal friends. John Polkinghorne, whom she states she greatly admired, was one of the few scientists who was a great encouragement to her, partly because he helped her realize that "atheism was not an essential prerequisite of science and not all scientists were as atheistic as they seemed."[523]

---

517 White and Gribbin, 1992, pp. 274-277.
518 *Hawking, 2004, p. 336.*
519 *White and Gribbin, 1992, p. 289.*
520 Hawking, 2004, p. 476.
521 Hawking, 2004, p. 484.
522 Hawking, 2004, p. 485.
523 Hawking, 2004, p. 246.

Jane's assessment of the religion-science conflict was especially insightful because she was able to stand back and objectively observe first-hand both worlds of science and religion, allowing her to make some objective judgments. Indeed, her writings in the 2004 book referenced here clearly represent an effort to come to grips with some of the central human questions, and why she accepted theism and rejected the atheism of virtually all the leading scientists including, especially, her husband, with whom she had spent much of her life. She was the proverbial fly on the wall, giving us insight, found nowhere else, into the thinking of the world's leading cosmologists.

The Enigma of Evil

Stephen's progressing illness caused endless hospital stays that required overcoming almost insurmountable obstacles to live a somewhat normal life. Evil was a subject with which Jane had to deal on a daily basis. The world God created provided motivation for discovery, and a freedom of choice provided a sense of wonder. Jane recognized that this freedom lies at the heart of the source of suffering and evil. God could eliminate evil, but if He did, freedom of choice also would be eliminated. She noted that if "belief in God were automatically decreed by the Creator, the human race would simply be a breed of automatons."[524] She stressed that most evil is often reducible "to human greed and selfishness."[525]

Stephen Abandons his Wife

Although many other women might have left Stephen because of his intolerable attitude toward her, and especially what she represented, Jane stuck by her husband through everything. It was he who left her for another woman. In short, Stephen Hawking "divorced his wife, Jane, primarily because she became a creationist."[526] She tried in vain to reconcile with Stephen—his terms included that he would live at home with his family for part of the week, and the rest of the week he would live "with his ladylove."[527] This condition was unacceptable to Jane, as it would be to most other Western women. His selfishness and hedonism had shown through again. Ironically, when asked "What do you think about during the day," Hawking said, "Women. They are

---

524 Hawking, 2004, p. 461.
525 Hawking, 2004, p. 461.
526 Mills, 2003, p. 119.
527 Hawking, 2004, p. 574.

a complete mystery."⁵²⁸

Much of her life is a contrast between a woman deeply conscious of her Christian spirituality, and a man firmly closed to any theistic considerations. Jane learned that faith is the outward expression of one's spirituality that "can make sense of all the wonders of Creation and of all the suffering in the world" and give "substance to all our hopes. However far-reaching our intelligent achievements, and however advanced our knowledge of Creation, without faith and a sense of our own spirituality there is only isolation and despair."⁵²⁹ When their split became public, the Cambridge academic community was shocked because for

> as long as anyone could remember, Stephen had taken great pains to promote the role Jane had played in his life and, despite their disagreements, to outsiders their marriage was a model of security. … There were rumors of extramarital relations developing over a number of years long before their marriage had reached crisis-point; but those who knew the couple well regarded as far more significant tales of increased tensions between Stephen and Jane over the old religious arguments. Their disagreements had been swept under the carpet for many years, but with the writing of *A Brief History of Time*, it appears that the wounds had been re-opened.⁵³⁰

One cannot read Jane's writings without admiring her and empathizing with the struggles that she faced daily with her militantly atheist husband. Her life provides an important lesson for all people interested, not only in the science/religion conflict, but also in understanding the basic human needs all of us possess.

Hawking Openly Argues for Atheism

Stephen Hawking makes it clear in his latest book about his "M-theory" he believes science has now eliminated the need for God to explain the universe.⁵³¹ Richard Dawkins said about the book, Hawking "finishes off God. Darwin kicked him out of biology, but physics remained more uncer-

---

528 DeLange, 2012, p. 27.
529 Hawking, 2004, p. 594.
530 White and Gribbin, 1992, p. 285.
531 Hawking and Mlodinow, 2010.

tain. Hawking is now administering the coup de grace" to God in physics.[532] And Hawking administered the *coup de grace* to God in physics by concluding a universe can spring into existence from nothing, a preposterous belief that negates the need for "some supernatural being or god."[533]

His critics say Hawking has done no such thing. Even some atheists have complained that Professor Hawking's book is "deficient in any meaty arguments that might begin to back up such weighty assertions [such as a universe springing into existence from nothing], and it provides few references to other books or articles that would provide this missing content."[534] Larsen wrote that the "mysterious, over-arching M-theory was enthusiastically embraced by many researchers" including Hawking who

> believed that not taking "this web of dualities as a sign we are on the right track would be a bit like believing that God put fossils into the rocks in order to mislead Darwin about the evolution of life." Despite its promise, M-theory was admittedly incomplete. Hawking described it as a jigsaw puzzle, in that it is "easy to identify and fit together the pieces around the edges but we don't have much idea of what happens in the middle," where the edges of the puzzle correspond to the various string theories and supergravity.[535]

Conclusions from Jane Hawking

Hawking's former wife has carefully documented the results of going from Christianity to atheism in the life of her famous scientist husband. She has shown how he has lived a selfish, immoral life in harmony with the beliefs that he has claimed are supported by scientific fact. His "science" is made up of some facts and much speculation, as shown by his latest book that disproves some of his former strongly held ideas. In my view, Stephen Hawking was an atheist because he does not want a God to exist. To justify his atheistic conclusions, Hawking claims that they are the result of objective scientific facts. However, little evidence exists to support his position.

The parallels between Darwin and Hawking are striking. For both of them, their life's work revolved around the attempt to explain reality without

---

532 Trottier, 2011, p. 55.
533 Trottier, 2011, p. 56.
534 Trottier, 2011, p. 55.
535 Larsen, 2007, p. 138.

God. Darwin tried to explain the origin of species (actually the origin of life, despite objections to the contrary by many evolutionists…) by a purely natural process, whereas Hawking tried to explain the origin of the universe by a totally natural process. Furthermore, both had devout Christian wives who were creationists. Lastly, both received much truculent criticism from their professional colleagues, Darwin from biologists,[536] and Hawking from cosmologists.

References

Anonymous. 2012. "Hawking at 70." *New Scientist*, pp. 26-27, January 7.

Bowler, Peter J. 2009. "Darwin's Originality." *Science*, 323:222-226, January 9.

De Lange, Catherine. 2012. "Stephen Hawking at 70." *New Scientist*. 213(2846):26-27. January 7.

Dowe, Phil. 2005. *Galileo, Darwin, and Hawking: The Interplay of Science, Reason, and Religion*. Grand Rapids, MI: William B. Eerdmans.

De Lange, Catherine. 2012. "Stephen Hawking at 70." *New Scientist*. 213(2846):26-27. January 7.

Hawking, Jane. 2004. *Music to Move the Stars*. New York: McMillan.

Hawking, Steven. 1988. *A Brief History of Time: From the Big Bang to Black Holes*. New York: Bantam.

_____. 2001. *The Universe in a Nutshell*. New York: Bantam.

Hawking, Steven and Leonard Mlodinow. 2010. *The Grand Design*. New York: Random House.

Larsen, Kristine. 2007. *Stephen Hawking: A Biography*. Amherst, NY: Prometheus Books.

Mills, David. 2003. *Atheist Universe; Why God Didn't Have a Thing to do with it*. Xlibris.

Schaefer, Henry F. 2003. *Science and Christianity: Conflict or Coherence?* Athens, GA: The University of Georgia.

Trottier, Justin. 2011. "A Little Too Grand?" *Skeptical Inquirer*. March/ April 2011. P. 55-56.

---

536 Bowler, 2009, p. 223.

White, Michael and John Gribbin. 1992. *Hawking; A Life in Science.* New York: Dutton.

Videos

Stephen Hawking's Grand Design. The Meaning of Life (36:32)
https://www.youtube.com/watch?v=usdqCexPQww

Ravi Zacharias Answers Stephen Hawking – Part 1. (18:27)
https://www.youtube.com/watch?v=0wMyMmjPgLs

Carl Sagan, Stephen Hawking and Arthur C. Clarke - *God, The Universe and Everything Else* (1988) https://www.youtube.com/watch?v=HKQQAv5svkk

The Origin of the Universe (1:07:24)
https://www.youtube.com/watch?v=c7MO3up5bZs

Chapter 28

Kevin R. Henke, Ph.D.

From YEC to Agnosticism

Dr. Kevin Henke was born into a family that regularly attended the United Methodist Church. In the early 1970s, Kevin and his younger brother both went through confirmation and joined the Beaver Crossing United Methodist Church. In 1975, he began attending a United Church of Christ College. In a philosophy class in this church-run college the professor that taught the class with the goal of causing students to distrust the Bible, teaching "it had its faults just like any other" book.[537] At this time, his college roommate happened to be a born-again Christian who got him involved in various religious activities. As a result, Henke professed to be "born again" in June of 1977.

Kevin then began attending what he now calls the campus fundamentalist group. He read the Bible in its entirety, much of which disturbed him, mostly the wars which were discussed in the Old Testament. Then, also in 1977, he became very interested in geology. He added that he was at that time "strongly opposed to biological evolution and was sympathetic and open to fundamentalist creationism with its young-Earth and Flood geology. I even defended fundamentalist creationism during a field project in Colorado in the summer of 1978."[538] After taking a couple of college science courses, he studied enough geology to conclude

> that the version of creationism taught by fundamentalists was bogus. I became an old-earth creationist, but kept my anti-evolution views until the mid-to-late 1980s. Although as an old-earth creationist I no longer took Genesis 1-11 and some other parts of the Bible literally, I still strongly believed in the inerrancy and infallibility of the Bible. Furthermore, I remained very impressed by the biblical creation account. To me it represented a generally "accurate" geological description of how the world was created, told in poetic fashion.[539]

---

537 Babinski, 1995, p. 244.
538 Babinski, 1995, p. 247.
539 Babinski, 1995, p. 247.

In May of 1979, he graduated with a B.A. in the physical sciences with minors in geology and physics.[540] Then in the summer of 1979, he spent much of his free time reading and memorizing Scripture. In August he moved to North Dakota to attend graduate school at the University of North Dakota to earn a M.S. in Geology. Once he arrived in North Dakota, he became very active in an on-campus chapter of Inter-Varsity Christian Fellowship.[541] He noted that some of his "fundamentalist acquaintances warned" him that he would completely abandon Christianity because

> by rejecting part of the Bible, I would soon reject the entire book and the Christian faith. One woman even suggested that I would be better off not reading "certain" books and magazines. I cringed at the thought of this "ignorance is bliss" attitude. What good is a faith that cannot stand up to simple questions and criticisms? Now, I certainly recognize that many liberals can reject parts of the Bible and retain a stable faith that will last for the rest of their lives. However, for me, this was not the case. The fundamentalists were right: my liberalism eventually evolved into disbelief. In the late 1980s, I began to read more literature that was skeptical of the virgin birth, the resurrection of Christ, and other Christian doctrines.[542]

Henke is now a former active YEC who actively opposes creationists of all stripes. He has a Ph.D. in Geology from the University of North Dakota, and for a while taught at the University of Kentucky. He currently is a Senior Research Scientist at the University of Kentucky's Center for Applied Energy Research.

Henke is well-known for his aggressive criticism of young-Earth creationism and scientific arguments used to document a young Earth. In particular, he has been very critical of ICR's RATE project's results, which claim to show evidence for a young Earth, such as the example that zircons contain too much helium to be billions of years old. He even has claimed that Russell Humphreys, a young-Earth creationist involved in the project, has made errors in his research. Humphreys stands by his conclusions.[543] Henke's publications include:

---

540 Babinski, 1995, p. 247.
541 Babinski, 1995, p. 247.
542 Babinski, 1995, p. 250.
543 Babinski 1995, pp. 240-252. In his chapter, he documents why he rejected creationism.

"Critical review of mercury contamination issues relevant to manometers at natural gas industry sites: GRI-93/0117, Gas Research Institute, Chicago (1993) with V. Kuhnel, R.H. Fraley, C.M. Robinson, D.S.

Charlton, H.M. Gust, and N.S. Bloom "The subsurface hydrology around Building 9201-2: Results of the July 1994 water level recovery test, Oak Ridge Y-12 Plant, Oak Ridge, Tennessee: Environmental Sciences Division, Oak Ridge National Laboratory, Y/ER-263 (1996) with M.A. Bogle and R.R. Turner.

"Structure and powder diffraction pattern of 2,4,6-Trimercapto-s-triazine, Trisodium salt (Na3s3C3N3-9H2O)," *Powder Diffraction* v.12, n.1, (March 1997) pp.7-12 with J.C. Bryan and M.P. Elless.

"Chemistry of heavy metal precipitates resulting from reactions with Thio-Red." 1998. *Water Environment Research*, 70(6):1178-1185, Sept-Oct.

"Chemical precipitation of mercury: Commercial claims and new approaches." 1999. Energeia, Center for Applied Energy Research, University of Kentucky with A.R. Hutchison, M.K. Krepps, M.M. Matlock & D.A. Atwood.

"The determination of the acid dissociation constants of Trithiocyanuric acid," *Abstracts and Program of 32nd American Chemical Society* Central Regional Meeting (May 1999) with A.R. Hutchison, J.T. Twyman, and D.A. Atwood.

Minerals in Thin Section. 2000. Prentice-Hall: Englewood Cliffs, NJ with D. Perkins.

"A pyridine-thio ligand with multiple bonding sites for heavy metal precipitation," *Journal of Hazardous Materials* B82, pp. 55-63 (2001) with M.M. Matlock, B.S. Howerton, and D.A. Atwood.

References

Babinski, Edward T. 1995. *Leaving the Fold: Testimonies of Former Fundamentalists*. Amherst, NY: Prometheus Books,

Video: Dr. Kevin Henke and Dr. Gary Loechelt discuss creationist claims about helium diffusion rates.
Part 1: (1:28:46) https://www.youtube.com/watch?v=fXD33E9c7pk
Part 2: (2:18:14) https://www.youtube.com/watch?v=HmhHLJaH-w0

Chapter 29

Denis Lamoureau, D.D.S., Ph.D.

From Atheist to Creationist to I. D. Supporter to Veneer Theist

Denis Lamoureau was once an atheist, then became a YEC, and then, due to what he claims was his reading and study of the scientific evidence, he became an Intelligent Design supporter. In the end, he finally rejected ID and became a theistic evolutionist. He considers himself to be an evangelical Christian. In his 2009 book titled, *I Love Jesus & I Accept Evolution*, he spills much ink supporting evolution with such arguments as the now-debunked claim that only a one percent genetic difference exists between chimps and humans and the now-refuted sketch of whale evolution, still treating these icons of evolution as if they were fully valid. He knows that abandoning the YEC view often gives parents difficulty in coping

> with such an upheaval in a child. "The day I had to tell my mother I wasn't a young-Earth creationist was the scariest day of my life," ... His mother was so embarrassed by his work in biology that she told her friends her son was still in the profession he once belonged to: dentistry. Some compare these conversations to informing fundamentalist Christian parents that they are gay—but perhaps even more wrenching.[544]

Lamoureau is now an Associate Professor of Science and Religion at the University of Alberta, Canada's St. Joseph's College, teaching his students about his belief of the fallacy of both creationism and ID. He holds three doctorates: one in dentistry, a second in theology, and a third in biology. He does not just attempt to prove evolution, but he also tries to create a rationale for the theistic evolution worldview. Lamoureau is now active in the American Scientific Affiliation, an organization once supportive of Biblical creation, but now devoted to theistic evolution. Many of his publications attempt to show that creationism and ID are both refuted by science fact and scripture. Note that many articles are in "Christian" science journals:

"Philosophy vs. Science" *Creation Science Dialogue* 8:3 (1981), 3.

---

[544] Couzin, 2008, p. 1036.

"Views on the origin of the universe and life," *What is Logos?* (1996) [2002, 3rd ed.], 37-40.

"A box or a black hole" A response to Michael J. Behe." *Canadian Catholic Review*, 17:3 (1999).

"Coming to terms with concordism and evolution." *Canadian Catholic Review*, 17:3 (1999).

"Charles Darwin and Intelligent Design" *Journal of Interdisciplinary Studies*, 15:1 (2003), 23-42.

"Evolutionary creation: Beyond the evolution vs. creation debate" *Crux: A Quarterly Journal of Christian Thought and Opinion*, 39:2 (2003), 14-22.

"Theological insights from Charles Darwin" *Perspectives on Science and Christian Faith* 56:1 (2004), 2-11.

"Lessons from the Heavens: On scripture, science and inerrancy" *Journal of the American Scientific Affiliation*, 60:1 (2008), 4-15.

"Robert Larmer on Intelligent Design: An evolutionary creationist critique" *Christian Scholars Review*, 37:1 (2007), 77-90.

"Gaps, design and theistic evolution: A counter response to Robert Larmer" *Christian Scholars Review*, 37:1 (2007), 101-116.

"The erosion of Biblical inerrancy, or toward a more Biblical view of the inerrant word of God?" *Perspectives on Science and Christian Faith*, 62:2 (2010), 133-138.

"Evolutionary creation: Moving beyond the evolution vs. creation debate" *Christian Higher Education* 9:1 (2010), 28-48.

"Was Adam a real person?" *Christian Higher Education*, 10:2 (2011), 79-96.

Some writings that refute Lamoureau's claims.

Behe, Michael J. *Darwin's Black Box; The Biochemical Challenge To Evolution.*

New York, NY: The Free Press, 1996, 307 pp. A best seller and one of the major texts of the intelligent design movement which argues that irreducible complexity renders the macroevolution paradigm impossible, at least in its early stages. Focuses specifically on the origin of life, the cell, and the major life forms. In this work he develops his famous mouse trap analogy arguing that a mouse trap is totally functionless unless it has a certain minimum number of parts, without which the trap cannot function, and thus is an irreducibly complex machine. Behe then argues that the cell and many of the structures in the cell are likewise irreducibly complex (only much more so than a simple mousetrap).

_____. 2007. *The Edge of Evolution*. New York: The Free Press.

_____, William A. Dembski and Stephen C. Meyer. *Science and Evidence for Design in the Universe*. San Francisco, CA: Ignatius Press, 2000, 234 pp.

Dembski, William A. (ed). *Mere Creation; Science, Faith & Intelligent Design*. Downers Grove, IL: InterVarsity Press, 1998, 475 pp. A discussion of the intelligent design movement by several of its leading proponents.

_____. *The Design Inference; Eliminating Chance through Small Probabilities*. Cambridge University Press, 1998, 244 pp. A discussion of the evidence for design and how this evidence can be evaluated.

_____. *No Free Lunch; Why Specified Complexity Cannot Be Purchased without Intelligence*. Lanham, MA: Rowman and Littlefield Publishers, Inc., 2002, 404 pp.

_____ (editor). *Darwin's Nemesis: Phillip Johnson and the Intelligent Design Movement*. Downers Grove, IL: Inter-Varsity Press, 2006, 357 pp.

_____ and James M. Kushiner (editors). *Signs of Intelligence; Understanding Intelligent Design*. Grand Rapids, MI: Brazo Press, a division of Baker Book House Co., 2001, 224 pp. A review of intelligent design theory for the general reader. The 14 essays included discuss the evidence and problems of the intelligent design world view and philosophy.

_____ and Jay Wesley Richards. *Unapolgetic Apologetics*. Downers Grove, IL: InterVarsity Press, 2001, 280 pp. Apologetics from an intelligent design worldview. Many of the authors are Princeton graduates and often refer to Christian work done at Princeton.

_____ and Michael Ruse (editors). *Debating Design; From Darwin to DNA.* NY: Cambridge University Press. 2004.

_____ and Sean McDowell. *Understanding Intelligent Design.* Eugene, OR: Harvest House, 2008. A basic review of ID for laypersons.

_____ and Thomas Schirrmacher (editors). *Tough-Minded Christianity.* Nashville, TN: B&H Publishing, 2008, 768 pp.

_____ and Michael R. Licona (editors). *Evidence for God.* Grand Rapids, MI: Baker Books, [indent]2010, 272 pp.

DeWolf, David K., Stephen C. Meyer and Mark E. DeForrest. *Intelligent Design in Public School Science Curricula: A Legal Guidebook.* Richardson, TX: Foundation for Thought and Ethics, 1999, 42 pp. Covers the legal case supporting teaching intelligent design.

Gordon, Bruce and William Dembski. 2011. *The Nature of Nature: Examining the Role of Naturalism in Science.* Wilmington, DE: Intercollegiate Studies Institute, 2011.

Meyer, Stephen. C. *Signature in the Cell.* New York: Harper One, 2009, 612 pp. A history of the Intelligent Design movement from the perspective of one of the leading advocates Cambridge educated, Meyer recounts his research there.

_____. *Darwin's Doubt: The Explosive Origin of Animal Life and the Case for Intelligent Design.* New York: Harper One, 2009, 478 pp.

Wells, Jonathan. *Icons of Evolution: Science or Myth?* Washington, DC: Regnery, 2002.

_____. *The Politically Incorrect Guide to Darwinism and Intelligent Design.* Washington, DC: Regnery Publishing, 2006, 273 pp.

_____. *The Myth of Junk DNA.* Seattle, WA: Discovery Institute Press, 2011, 174 pp.

Reference

Couzin, Jennifer. 2008. "Evolution: Crossing the Divide." Science, 319:1034-1036, February 22.

Videos

Struggling With Origins (17:33)
https://www.youtube.com/watch?v=5t4-ti00lrg

Beyond the Evolution vs. Creation Debate (9:25)
https://www.youtube.com/watch?v=352HUa5sBn0&list=PL40F268519087C43F

Was Adam a Real Person?
Part 1 (9:06)
https://www.youtube.com/watch?v=BLzv9xrdNOs&list=PL0DA1C289C6B8D0B1
Part 2 (9:07)
https://www.youtube.com/watch?v=oMbKKtbJNr0
Talks about his book Evolutionary Creation and why he believes Adam never existed.

The Evolution of an Evolutionary Creationist Part 1 (14:29)
https://www.youtube.com/watch?v=PD2i7tD4qqA

Beyond the "creation vs. evolution" Debate (14:13)
https://www.youtube.com/watch?v=QaeGfV-N2kM

Chapter 30

Louis Seymour Bazett Leakey, Ph.D.

From Aspiring Minister to Militant Darwinist

The Leakey family members are the world's most accomplished paleoanthropologists in history. Louis Leakey's conversion from aspiring minister to militant opponent of orthodox Christianity is yet another excellent example of the adverse influence of Darwinism on not only Christianity, but also theism in general. As a young man, Louis, the son of missionaries, was so "zealous about his Christianity" that he "sometimes stood on corner soap boxes to deliver sermons."[545] One reason for his enthusiasm for Christianity was that he saw firsthand some of the many clear benefits that Christianity conferred on the African tribes where he grew up.[546] As a Cambridge University student, he even chastised some of his fellow students for not "being proper Christians."[547]

His early ambition was to be a missionary and, until 1925, he decided that he could be "both a missionary and a part-time scientist."[548] As Leakey studied at Cambridge, though, his "growing knowledge of evolutionary theory" led him away from theism. In his words, he became "firmly convinced of the truth of the theory of Evolution as distinct from Creation as described in Genesis."[549] He grew to feel that his new "scientific" views were no longer "compatible with missionary work," so he decided to pursue a full-time career in science.[550]

Leakey eventually became very hostile toward Christianity— and he passed on that attitude to at least one of his sons, Richard. When Richard was asked to be a guest on Walter Cronkite's television program as an "ardent anti-creationist, he agreed to go on" the show.[551] The request was a ruse to get him on the set—Cronkite wanted to pit Leakey and fellow Darwinist Donald Johanson against each other in order to force them to debate their radically dif-

---

545 Morell, 1995, p. 28.
546 Louis Leakey, 1966, pp. 386–388.
547 Morell, 1995, p. 28.
548 Morell, 1995, p. 33.
549 Louis Leakey, 1966, p. 161.
550 Louis Leakey, 1966, p. 161.
551 Morell, 1995, p. 520.

ferent opinions about *Australopithecus afarensis* and other putative hominids.

On the show, Johanson was less interested in an intellectual exchange to help him arrive at a better understanding of human evolution than in attacking those with whom he disagreed. Some feel that Richard Leakey came out much better in this exchange. Other people felt otherwise because shortly "after the Cronkite show, the National Geographic Society— the Leakeys' long and trusted supporter"—turned down Richard's request for support for new human fossil explorations north and west of Lake Turkana.[552]

The antagonism towards creationism by the Leakeys was so great that Mary Leakey "refused to send any of Kenya's original hominid fossils for display" at the American Museum of Natural History in New York because "such an exhibit was too risky in a country where creationists were active."[553] She believed that the fossils were "in danger" in a country where there were many creationists because she was afraid that the creationists would "come in with a bomb and destroy the whole legacy" of "irreplaceable fossils."[554] Of course, this belief is irresponsible because creationist paleontologists would love to carefully study the original fossils in detail.

References

Leakey, Louis 1966. *White African*, Cambridge, MA: Schenkman.

Morell, Virginia. 1915. *Ancestral Passions. The Leakey Family and the Quest for Mankind's Beginnings*. New York: Simon & Schuster.

---

[552] Morell, 1995, p. 523.
[553] Morell, 1995, p. 533.
[554] Morell, 1995, p. 533.

Chapter 31

Glenn Robert Morton

From Active YEC to YEC Opponent

Glenn was once an active YEC, publishing numerous articles in creation journals. He is now an active opponent of creationism, publishing many articles against his former worldview. Morton documents in detail why he rejected creationism on a website devoted to cases like his.[555] In short, after graduating from college with a degree in physics, he found employment as a geophysicist working for a seismic company. This is where he

> first became exposed to the problems geology presented to the idea of a global flood. I would see extremely thick (30,000 feet) sedimentary layers. One could follow these beds from the surface down to those depths where they were covered by vast thicknesses of sediment. I would see buried mountains which had experienced thousands of feet of erosion, which required time.[556]

He reasoned that

> the sediments in those mountains had to have been deposited by the flood, if it was true. I would see faults that were active early but not late and faults that were active late but not early. I would see karsts and sinkholes (limestone erosion) which occurred during the middle of the sedimentary column (supposedly during the middle of the flood) yet the flood waters would have been saturated in limestone and incapable of dissolving lime. It became clear that more time was needed than the global flood would allow.[557]

He worked for the next few years attempting to solve these problems. To achieve this goal, he had

> published 20+ items in the *Creation Research Society Quarterly*. I would listen to ICR, have discussions with people like Slusher,

---

555 Morton, 2000.
556 Morton, 2000.
557 Morton, 2000.

> Gish, Austin, Barnes and also discuss things with some of their graduates that I had hired.

In order to research the data more carefully "with the hope of finding a solution," in 1980 he left the seismic processing field and moved into seismic interpretation where he had

> to deal with more geologic data. My horror at what I was seeing only increased. There was a major problem; the data I was seeing at work, was not agreeing with what I had been taught as a Christian. Doubts about what I was writing and teaching began to grow. Unfortunately, my fellow young earth creationists were not willing to listen to the problems. No one could give me a model which allowed me to unite into one cloth what I believed on Sunday and what I was forced to believe by the data Monday through Friday. I was living the life of a double-minded man--believing two things.[558]

In short, by 1986, according to his post on *no answers in Genesis* website. His

> growing doubts about the ability of the widely-accepted creationist viewpoints to explain the geologic data led to a nearly 10-year withdrawal from publication. My last young-earth paper was entitled Geologic Challenges to a Young-earth, which I presented as the first paper in the First International Conference on Creationism. It was not well received.[559]

Notice that the basic issue, as was true of many who rejected YEC, was first the age of the Earth, then the universal Flood, and, last, came the acceptance of Darwinism. He now writes articles such as "The Imminent Demise of Evolution: The Longest Running Falsehood in Creationism." In this article, he gave numerous examples of these failed predictions dating back to 1825.[560] While not an atheist, he is active in refuting the major arguments for theism and his work is used by atheists to bolster their case. Morton is now active in the *American Scientific Affiliation* publishing articles to discredit both creationism and ID. His many publications include the following:
"Can the canopy hold water?" *CRSQ* 16(3): 164-169, December 1979.

---

558 Morton, 2000.
559 Morton, 2000.
560 Morton, 2010.

"The warm earth fallacy" *CRSQ* 17(1): 40-41, June 1980.

"Prolegomena to the study of the sediments" CRSQ 17(3): 162-167, December 1980.

"Creationism and continental drift" *CRSQ* 18(1): 42-45, June 1981.

"Electromagnetics and the appearance of age" *CRSQ* 18(4): 227-232, March 1982.

"Fossil succession" *CRSQ* 19(2): 103-111, 90, September 1982.

"The Flood on an expanding Earth" *CRSQ* 19(4): 219-224, March 1983.

"The age of lunar craters" *CRSQ* 20(2): 105-108, September 1983 with Harold S. Slusher and Richard E. Mandock.

"The carbon problem" *CRSQ* 20(4): 212-219, March 1984.

"The age of oil and gas" *CRSQ* 20(4): 229- 230, March 1984.

"Global, continental and regional sedimentation systems and their implications" *CRSQ* 21(1): 23-33, June 1984.

The Geology of the Flood. DMD Publishing Company: Dallas, Texas, 1985.

"Geologic challenges to a young Earth" *Proceedings of the First International Conference on Creationism* Volume II: Technical Symposium Sessions and Additional Topics (Creation Science Foundation: Pittsburgh, PA, 1986), pp.137-145.

"Mountain synthesis on an expanding Earth" *CRSQ* 24(2): 53-61, September 1987.

"Changing constants and the cosmos" *CRSQ* 27(2): 60-67, September 1990.

Foundation, Fall and Flood: A Harmonization of Genesis and Science DMD Publishing Company: Dallas, Texas, 1995.

References

Morton, Glenn R. 2000a. "Why I left Young-Earth Creationism." *Old Earth Ministries* (blog) http://www.oldearth.org/whyileft.htm

_____. 2000b. The Transformation of a Young-earth Creationist. Perspectives on Science and Christian Faith. 52(2):81-83.

_____. 2010. "The Imminent Demise of Evolution: The Longest Running

Falsehood in Creationism." *The Word of Me* (blog). April 3. https://thewordofme.wordpress.com/2010/04/03/the-imminent-demise-of-evolution-the-longest-running-falsehood-in-creationism/#comment-44993

Chapter 32

David Mills

From Devout Christian to Devout Evolutionist

Another person who left his Christian faith because of Darwinism is the best-selling author David Mills. In the early 1970s he described himself as a devout believer, "absolutely convinced that Jesus' second coming was at hand."[561] He professed to be "saved" and was baptized when he was only nine years old. He regularly attended the Baptist church for the next sixteen years. During his first year of high school he

> began associating with some friends who were ... exuberant about their "relationship with the Lord." Their enthusiasm soon rubbed off on me. I myself became a glassy-eyed religious fanatic, parading around high school distributing pamphlets titled *What Must I do to be Saved?* I preached the Gospel to any lost souls who would listen: and I felt deep satisfaction whenever someone called me a "Jesus freak," since I considered that label to be a badge of high honor.[562]

Then he encountered Darwinism and his life soon changed drastically. As Mills explains in his book *Atheist Universe: The Thinking Person's Answer to Christian Fundamentalism*, the more he learned about evolution, the more disenchanted he became with creation science and Christian logic. His book focuses on what he calls the irrationality of Christian belief and the Bible teaching of a young earth, Noah's Flood, and the Genesis creation account. Learning about evolution was not really learning, but exposure to a very biased presentation, either by a professor or TV program. When creationists get a fair fight with Darwinians, creationists win and this is why it is so difficult to set up creation v. evolution debates with Darwinists. The apostasy occurs when people are exposed to biased, Darwin-Only propaganda sessions.

Ironically, most all of his arguments have been refuted long ago by a variety of scholars.[563] He concluded the "language used by Christian-apologist

---

561 David Mills, 2003, pp. 73-74.
562 David Mills, 2003, p. 74.
563 See *How We Got the Bible* by Neil R. Lightfoot. Grand Rapids, MI: Baker Books, 2003, Third Edition. Also see *Evidence That Demands a Verdict: Life-Changing Truth for a Skeptical World* by Josh

writers is *deliberately* obscure and jargon-filled to create the *façade* of intellectual respectability."[564] After making these claims, he did not immediately become an atheist, but still believed that Christianity could be accepted by *faith*, and not because of the scientific evidence, because he felt none existed. As a result,

> the absence of evidence to *support* Christianity had little effect on my beliefs. What finally turned me toward atheism was my realization that science not only could not *confirm* Christian teachings, but offered powerful evidence *against* the Bible as well. For example, the Genesis accounts of Creation, Noah's flood, and the age of the Earth are provably false, as are numerous other Old and New Testament fables.[565]

Mills eventually decided he had no basis to accept Christianity on faith alone without evidence. He is now very active preaching his Darwinism, and indirectly atheism, on the radio, in print, and wherever he has an opportunity.

References

Lightfoot, Neil R. 2003. *How We Got the Bible*. Grand Rapids, MI: Baker Books, Third Edition.

McDowell, Josh with Sean McDowell, Ph.D. 2017. *Evidence That Demands a Verdict: Life-Changing Truth for a Skeptical World*. Nashville, TN: Thomas Nelson Publishers.

Mills, David. 2003. Atheist Universe; *Why God Didn't Have a Thing to do with it*. Bloomington, IN: Xlibris.

McDowell with Sean McDowell, Ph.D., 2017, Nashville, TN: Thomas Nelson Publishers.
564 David Mills, 2003, p. 75.
565 David Mills, 2003, p. 76.

Chapter 33

P.Z. Myers, Ph.D.

From Christian to Militant Christian-Hater

Paul Zachary (PZ) Myers, born March 9, 1957, is now an Associate Professor of Biology at the University of Minnesota, Morris. Myers was reared in an Evangelical Lutheran church. Prior to his confirmation, Myers says, "I started thinking, you know, I don't believe a word of this." He did not provide any details of his conversion to atheism except he claims that it was not due to a bad experience in the church. Now a militantly aggressive atheist, Myers comments widely on the blog he founded (titled Pharyngula) about science, education, atheism and religion. Pharyngula is currently hosted on both the Science and Free Thought Blog networks. In 2006, based on his popularity, the journal Nature listed his Pharyngula web-site as the top-ranked blog hosted by a scientist.[566] This blog presents a no-holds barred, vitriolic condemnation of both creation and Intelligent Design (ID). Many of the blog posters freely use the most obscene language possible, and are proud of their reliance on scatological language that would make a sailor blush.

Myers claims that he currently works with zebrafish in the evolutionary developmental biology field, but I was not able to find much that he has published in this area.[567] As an outspoken critic of ID and of all branches of the entire creationist movement, he is very active in the American creation–evolution controversy. An aggressive, confrontational person, Myers says he has "nothing but contempt" for Intelligent Design, calling it "fundamentally dishonest."

Myers received the American Humanist Association's 2009 Humanist of the Year Award and the International Humanist Award in 2011. Asteroid "153298 Paulmyers" was also named in honor of him. He is now among the world's most popular atheist speakers, named one of the "Four Horsemen," of atheism along with Richard Dawkins, Sam Harris, and the late Christopher Hitchens. Myers' emotionalism has helped him to earn the label of having an

---

566 "Top five science blogs," Nature, 442 (7098): 9–9, 6 July 2006.
567 Bergman, Jerry, 2016. *Silencing the Darwin Skeptics*, (Volume II in the Slaughter of the Dissidents trilogy) Leafcutter Press. Take a look at chapter 8 (Nathaniel Abraham: an advocate of creation who did research on zebrafish). He was terminated in 2004 from Woods Hole Oceanographic Institute simply because he refused to personally adhere to the concept of evolution. This belief had no impact on his ability to do the research he was working on at the time.

"eccentric voice" by *USA Today*. In a 2014 lecture, Myers said he is labeled a "weird," "loud," "radicalized" and a "firebrand" type of "New Atheist."

Video: *Science Education: Caught in the Middle in the War between Science and Religion* (1:19:21) https://www.youtube.com/watch?v=R0yxh85xRCY

Video: *Facts, Theories, and Meaning: Evolution is more Complicated Than You Think*. (57:37) River City Reasonfest in Winnipeg on September 20, 2015. https://www.youtube.com/watch?v=oF7DQUaTgCA

Video: *The Crucifixion was Nonsense* (1:53:39)
https://www.youtube.com/watch?v=BLusSLLywyA

Video: *Bad Biology – How Adaptationist Thinking Corrupts Science* (1:20:28) https://www.youtube.com/watch?v=QVZb1B36TDg&feature=youtu.be

Video: *The Inescapable Conflict Between Science and Religion* (50:54)
From the Global Atheist Convention – MCEC Melbourne, March 2010.
https://www.youtube.com/watch?v=qfXbnygJirc
Myers claims that "science and religion are incompatible.".

Video: *Science and Atheism: Natural Allies* (55:29)
https://www.youtube.com/watch?v=3U0MnBmSlhE&feature=youtu.be
Myers states that "atheism and science belong together."

Video: *Design vs. Chance* (46:20)
At the 2009 Atheist Alliance International conference in Burbank, CA.
https://www.youtube.com/watch?v=ba2h9tqNYAo&feature=youtu.be

After setting himself up as an expert on Intelligent Design (ID) based on attending lectures and reading papers by some of the foremost ID advocates, Myers then launches into an incredibly lame diatribe. He states that ID arguments can be summed up in one word: "complexity, complexity, complexity." In doing this, Myers provides a very inaccurate view of what most ID proponents actually say about ID tenets. Myers' unfair characterization is akin to criticizing the effectiveness of car tires without mentioning anything about the need for a tread pattern. Furthermore, to characterize all tires without mentioning the benefits of tread is a serious omission. This is essentially the major flaw of Myers' reasoning in this video as he fails to accurately represent the key aspects of ID arguments concerning complexity. His attempt to invalidate ID propositions by misrepresenting them is an unfortunate stance, and should cause us to question his approach in other arguments he puts forward that are critical of ID. His omission of the well-established arguments for complexity in the ID literature is so blatantly absent that I could not let it stand without comment here.

In this video PZ Myers sets up a straw man argument. I can only conclude that his presentation in this video is deliberately misleading because he knows full well what the ID arguments for complexity are, and he has clearly misrepresented them here (for example, he has gone on record talking about *specified* and *irreducible* complexity).[568] But it's not unusual to see many critics of intelligent design appeal to their audience based on misleading representations and half-truths.

After claiming that the entire argument of intelligent design can be summed up with the words "complexity, complexity, complexity," he then seeks to mock this alleged fallacy as a simpleton's fairytale. One of the examples he uses to illustrate his point is a wall of driftwood logs strewn together on a beach. He cites this as evidence of complexity but then points out of course we all know that nobody designed the arrangement of this jumble of logs. Therefore, he claims that complexity needs no designer. It all happened naturally and without intelligent intervention. He then thinks that by providing this example of complexity he has laid bare the ridiculous nature of the whole intelligent design argument. Meyers is deliberately misleading his audience because he clearly leaves out at least two key ingredients of the ID proposition. And these are elements I know he is aware of, so it's a mystery why he fails to mention them early on in this video. He states that "… you can't use simple complexity as a measure of design"(13:50). But here Myers' argument is like suggesting that tires don't need tread. Mainstream ID advocates do not make claims about "simple complexity." They almost always refer to *specified* or *irreducible* complexity, therefore, Myers is constructing a straw man argument, and he knows better.

The *specificity* argument[569] posits that the evidence of designed things indicates a purpose, which is revealed in patterns of *information*. A jumbled pile of driftwood does not show us any pattern of information with a purpose. A flagellum motor clearly indicates specified complexity and purpose in all its parts. The flagellum motor exhibits information[570] because it is composed

---

568 https://www.youtube.com/watch?v=hpEhVvALhUM
569 Dembski, William. "Explaining Specified Complexity,"Metanexus, 13 September 1999. https://www.metanexus.net/explaining-specified-complexity/
570 Francis Crick defines information as "…the specification of the amino acid sequence of the protein." on p. 144 in "On Protein Synthesis" (1958), pp.138-163. https://profiles.nlm.nih.gov/ps/access/scbbzy.

of parts designed for the purpose of rotating the flagellum. Amino acid sequences in cell proteins also contain specified complexity, and this precise and repeated sequencing amounts to purposeful patterns not found in naturally occurring artifacts that Myers cites as "complex." There is no such purpose or information evident in a "complex" pile of driftwood. The specificity of complex functions evident in living systems has been well known to scientists for quite some time now. Also, specified complexity clearly implies an intelligent cause, not a naturalistic one. Anyone who can appreciate the obvious sees this distinctive nuance except for critics like PZ Myers. Furthermore, the source of specified complex information is easily attributed to an "intelligent source," even by many evolutionary advocates.[571] [572]

> 'A major enigma in evolutionary biology is that new forms or functions often require the concerted efforts of several independent genetic changes. It is unclear how such changes might accumulate when they are likely to be deleterious individually and be lost by selective pressure' [573]

Put another way, a string of letters could be considered by some to be complex because each letter has a unique pattern. This line of reasoning suggests that a string of random letters can be arranged to convey a level of complexity:

Ahjkfioprjpboirojkgajdlpyubnsmvlwtpnbbdpioebnhklyuaiotnwojf

This collection of randomly generated characters roughly parallels the type of complexity offered by Dr, Myers in his driftwood example. But this type of "simple" complexity is not what ID supporters are advocating. The notion of specified complexity is more appropriately illustrated when letters are formulated in such a way that they communicate information, such as:

The quick brown fox jumped over the lazy dog

---

pdf
571 Meyer, Stephen C.,"DNA and the Origin of Life: Information, Specification, and Explanation" p. 31. www.vedicilluminations.com/downloads/Intelligent-Design/DNAPerspectives.pdf From the abstract: "the term 'information' can designate several theoretically distinct concepts. By distinguishing between specified and unspecified information, this essay seeks to eliminate definitional ambiguity associated with the term 'information' as used in biology."
572 Woodmorappe, John. "Irreducible Complexity: Some Candid Admissions by Evolutionists." *Journal of Creation*, Vol. 17, No. 2 (August 1, 2003), pp. 56-59.
573 True, H.L. and Lindquist, S.L., "A yeast prion provides a mechanism for genetic variation and phenotypic diversity," *Nature* 407(6803):477–483, 28 September 2000, p. 477.

Clearly, information (like that found in biological systems) based on specified complexity is far different than Myers' false claim that ID arguments rely on "simple complexity."

The *irreducible* complexity argument is described by the creator of this term (Michael Behe) as referencing "...a single system which is composed of several interacting parts...where the removal of any one of the parts causes the system to cease functioning."[574] This notion clearly refers to a type of complexity that is anything but "simple" as proclaimed by Dr. Myer. Behe's concept envisions systems composed of coordinated parts that are required to cooperatively support specific functionality. And the upshot of this argument is that a naturalistic evolutionary approach is not sufficient to demonstrate the formation of such systems, since the likelihood that all the components required to form such systems coming together at the same time is highly improbable. In other words, irreducibly complex and coordinated systems and their functions exhibit a strong likelihood of intelligent design. Nevertheless, critics of ID often make statements like "it is not all that hard to imagine how eyes *might* have evolved by natural selection."[575]

It's worth pointing out that defending the alleged evolution of any system based on imagined scenarios (a very common tactic among many evolutionists that I refer to as Probability Theology) of what "might" have occurred via natural selection is hardly a more compelling counter-argument to irreducible complexity. The ubiquitous use of terminology such as "are thought to occur,"[576] "most likely evolved,"[577] "are presumed to be," "suggests that," "appear to have been," and so on are an unconvincing and non-scientific substitution for the evidence evolutionists do not have in hand, and we are under no obligation to accept such conjectures as fully reliable (like they do). This sleight-of-hand is used ubiquitously throughout the scientific literature to promote the reliability of evolution. This type of reasoning is so widespread among evolutionary biologists and paleontologists that it has become, in ef-

---

574 Behe, Michael. *Evidence for Intelligent Design from Biochemistry*. Speech presented to the Discovery Institute on August 10, 1996, http://www.arn.org/docs/behe/mb_idfrombiochemistry.htm
575 Irreducible complexity. No date, no author. http://content.csbs.utah.edu/~rogers/evidevolcrs/ircomp/index.html
576 Ohta, T., Near-neutrality in evolution of genes and gene regulation, *Proceedings of the National Academy of Sciences of the USA* 99 (25):16134–16137, 10 December 2002, p. 16136.
577 Carroll, Robert L. *Vertebrate Paleontology and Evolution*. NY: W.H. Freeman, 1988. http://doc.rero.ch/record/200124/files/PAL_E3902.pdf - This massive survey of fossil vertebrates is riddled with tons of admissions pointing to species stasis (ie, non-change) rather than evolution. It also illustrates the heavy use of terminology indicating that the evolution of most critters is largely based on speculation. Anyone who claims that the fossil record has tons of transitional fossils has not read this book.

fect, one of the most blatant articles of faith they promote wherever evidence is lacking. This type of thinking is what has unfortunately become the defacto basis for much of what is passed off as science in many evolutionary arguments we hear about today. Once we understand that this type of reasoning is the foundation that evolutionists build their arguments upon, we begin to see just how unreliable the evolutionary edifice really is.

I would direct readers to the following comment by Stephen Meyer of the Discovery Institute:

> From the abstract: "…the term 'information' can designate several theoretically distinct concepts. By distinguishing between specified and unspecified information, this essay seeks to eliminate definitional ambiguity associated with the term 'information' as used in biology."[578]

These two distinctions about the *type* of complexity as defined a prominent advocate of ID are clearly articulated in many ID books and videos, but Myers ignores them in this video by failing to adequately or fairly examine them. His argument is about as compelling as the claim that evolution can be summed up in one word: "change, change, change." As anyone who is even moderately familiar with evolutionary biology knows, this is hardly a fair description. So too is Meyers' straw man that ID is all about "[simple] complexity." It's distorted presentations like this that provide ample reason to remain skeptical of the obstinate scoffaholic characterizations of ID propositions coming from evo-partisan critics of ID.   - K. Wirth, Editor

---

[578] Meyer, Stephen C., "DNA and the Origin of Life: Information, Specification, and Explanation" [date unknown] 44 pp. www.vedicilluminations.com/downloads/Intelligent-Design/DNAPerspectives.pdf

Chapter 34

Ronald Numbers, Ph.D

Minister's Son and Christian to Agnostic

Noted University of Wisconsin historian of science, Ron Numbers, abandoned his belief in Christianity due to his exposure to Darwinian evolution. His father was a Seventh-Day Adventist (SDA) minister and, as a result, he learned about creationism very early in life, and even attended SDA church schools from first grade through college.[579] Both of his grandfathers, his father, and all his uncles on both sides worked for the church. His maternal grandfather was President of the International Church. His brother-in-law and his nephew are ministers.[580]

Things changed considerably when he began attending the University of California, Berkeley as a graduate student to study science. Due to his exposure to evolution there, he began to seriously doubt his religious faith, first by questioning the young-Earth teaching. He says that until then, he does

> not recall ever doubting the recent appearance of life on the earth until the late 1960s, while studying the history of science at the University of California at Berkeley. I vividly remember the evening I attended an illustrated lecture on the famous sequence of fossil forests in Yellowstone National Park and then stayed up much of the night with a biologist friend of like mind, Joe Willey, first agonizing over, then finally accepting, the disturbing likelihood that the earth was at least thirty thousand years old.[581]

He adds that what stands out in his

> memory as being decisive was hearing a lecture about the fossil forest of Yellowstone, given by a creationist who'd just been out there to visit. He found that for the 30 successive layers you needed—assuming the most rapid rates of decomposition of lava into the soil and the most rapid rates of growth for the trees that came back in that area—at least 20,000 to 30,000 years. The only alternative

---

579 Numbers, 2006, p. 13.
580 Paulson, 2007, p. 3.
581 Numbers, 2006, p. 13.

the creationists had to offer was that during the year of Noah's flood, these whole stands of forest trees came floating in, one on top of another, until you had about 30 stacked up. And that truly seemed incredible to me. Just trying to visualize what that had been like during the year of Noah's flood made me smile.[582]

He then "decided to follow science rather than Scripture on the subject of origins" and "quickly, though not painlessly, slid down the proverbial slippery slope toward unbelief."[583] He added that by 1982

> when attorneys for both sides in the Louisiana creation-evolution trial requested my services as a possible expert witness, I elected to join the ACLU team in defending the constitutional wall separating church and state. In taking my pretrial deposition, Wendell R. Bird, the creationist lawyer who had tried to recruit me for his side… publicly labeled me an "Agnostic." The tag still feels foreign and uncomfortable, but it accurately reflects my theological uncertainty.[584]

In one interview, Numbers outlined with stark clarity the theological implications that theistic evolutionists often do their best to evade. According to Christianity, said Numbers,

> humans were perfect because we were created in the image of God. And then there was the fall. Death appears and the whole account [in the Bible] becomes one of deterioration and degeneration. So we then have Jesus in the New Testament, who promises redemption. Evolution completely flips that. With evolution, you don't start out with anything perfect…. There's no perfect state from which to fall. This makes the whole [Christian] plan of salvation silly because there never was a fall.[585]

Contemporary theistic evolutionism—perhaps more accurately described as theistic *Darwinism*—is often presented to the public as a simple common-sense solution to the conflict between Darwin and Christianity. But, as the experience of Ron Numbers suggests, the problems raised by Darwin-

---

582 Paulson, 2007, p. 3.
583 Evans, 2009.
584 Numbers, 2006, p. 13.
585 Evans, 2009.

ism for Christians are not so easily assuaged. Wistfully, he remembers that he

> really wanted to have religious beliefs for a long time. I miss not having the certainty of religious knowledge that I grew up with. But after a number of years of trying to solve these issues, I decided they're not resolvable. So I think the term "agnostic" would be best for me.[586]

Ironically, the very Yellowstone forests that had pulled Numbers away from his faith have been re-interpreted due to research on Mt. St Helens as catastrophic mudflow deposits just like Yellowstone. This is another good example of evidence that swayed someone but was later reinterpreted. I've noticed that person after person gets persuaded away from their Christian faith by evidence that is later explained as false or non-compelling with regard to evolution, but they don't change their views. The work of Ernst Haeckel, especially his fraudulent comparative embryo drawings, is perhaps the best example of how fraud continued to influence thousands of people for decades, even long after his work depicting embryos was exposed as deliberate deception. In cases where fraud is not involved, a better understanding of arguments that really have so solid basis for explaining evolution, such as vestigial organs and the positive impacts of mutations, continues to this day to influence many people.

References

Evans, Gwen 2009, "Reason or Faith? Darwin Expert Reflects," *Wisconsin Week*, Feb. 3 .http://www.news.wisc.edu/16176

Numbers, Ronald L. 2006. *The Creationists: From Scientific Creationism to Intelligent Design*. [The Expanded Edition]. Cambridge, MA: Harvard University Press.

Paulson, Steve. 2007. Interview of Ron Numbers: "Seeing the Light of Science." http://www.salon.com/books/int/2007/01/02/numbers/.

Videos

Social Darwinism: A Historical Myth? (1:16:49)
ttps://www.youtube.com/watch?v=BliF1xcToPk

---

586 Paulson, 2007, p. 5.

The historical relationship between Science and Religion – 2017 (4:09)
https://www.youtube.com/watch?v=jcq8wcSbvJ4

Why Darwin Sill Matters (1:03:35) Pepperdine University (2009)
Talks about the worldwide impact of creationism and ID.
https://www.youtube.com/watch?v=XkuospzdlPY&list=PL93Y4RKvxDuw_gSGMMS399-qd3uYOy-Km

Chapter 35

George Perdikis

Creationist to Atheist Evolutionist

George Perdikis is the co-founder of the hugely successful Christian music band called Newsboys, one of the most popular Christian bands in history. In a guest post published at the Friendly Atheist's Patheos blog, Perdikis wrote that by the year 2007 he had "renounced Christianity once and for all and declared myself an atheist." His leaving Christianity produced a lot of negative press for Christians and a lot of favorable press for the atheists. He stated that one reason why he left Christianity was because he had developed an interest in cosmology and science.

He no doubt was influenced by the fact that most leading secular cosmologists are also fervent Darwinians, which is strongly reflected in their writing. As a result of this influence, he became "fascinated by the works of evolutionists and anti-creationists Carl Sagan, Neil deGrasse Tyson, Lawrence Krauss, Brian Cox, and Richard Dawkins."

He explains that his perceptions of Christianity began "to transform when I became interested in cosmology in 1992. …. I learned so much and was blown away by all the amazing scientific discoveries and facts" of science. He attributes the influence of many well-known evolution advocates such as Carl Sagan, Neil deGrasse Tyson and Richard Dawkins among others as being key influencers in his quest for understanding.[587] Perdikis explained it was his decision to become an atheist that caused him to leave the band in 1990 and to focus on other interests. He returned to his hometown Adelaide, Australia, got married, had two daughters, then his marriage 'dissolved.' After his marriage ended, he studied human psychology and evolution, and became an active evolutionist and an atheist.[588] This response by Perdikis to Darwinism's influence may have been one reason why the Newsboys band was a central part of the highly successful box office hit film titled *God is Not Dead*[589] which presents the evidence against evolution and for creation.[590]

---

587 Perdikis, 2015.
588 Yapching. 2015.
589 Watch the main song sung by the Newsboys from the movie soundtrack here: https://www.youtube.com/watch?v=FlxcIW4qd5w
590 See also the book *God's Not Dead: Evidence for God in an Age of Uncertainty* by Rice Brooks. Nashville, TN: Thomas Nelson Publishers, 2015.

References

Perdikis, George. 2015. "I Co-Founded One of the Most Popular Christian Rock Bands Ever… and Now I'm an Atheist." *Friendly Atheist*. January 21.

Yapching, Mark. 2015. "Newsboys co-founder George Perdikis has 'renounced Christianity once and for all'" 26 January. http://www.christiantoday.com/article/newsboys.founding.member.says.he.has.renounced.christianity.once.and.for.all/46698.htm

Chapter 36

William Provine, Ph.D.

From Creationist to Militant Atheist

The late Cornell Professor of Biology, William B. Provine, a former creationist, was reared by a devout Presbyterian minister father in what he describes as a "loving family." He became an atheist and an open opponent of Christianity after studying Neo-Darwinism and listening carefully to his evolution professor, Dr. Throckmorton, when a student at the University of Chicago. Provine recalled that, as a result of Throckmorton's presentation of Darwinian evolution as a blind, aimless process, he lost "for good" his belief that design exists in nature, writing that:

> Evolution exhibited no sign whatever of purpose. Evolution just happens. I can remember that the pain of loss [of belief in God] lasted less than a week. As the creationists claim, belief in modern evolution makes atheists of people. One can have a religious view that is compatible with evolution only *if the religious view is indistinguishable from atheism.*[591]

Yet the university and courts did not interfere with Provine's indoctrination, even though he had not only openly "expressed his religious viewpoint," by his own admission.

Until his death in 2015, Provine actively indoctrinated his students into evolution and atheism, and boasted that he was very successful in this endeavor. His method was to present the theistic position in his class, then, for the rest of the quarter, endeavor to demolish all of the arguments for theism. He used to brag that, at the beginning of his Cornell course, about 75% of his students were either creationists, or at least believed in purposeful evolution (i.e., were theists who believed that God directed evolution), but after the course, the percentage of theists had dropped to around 50 percent. For comparison, about 90 percent in society as a whole are theists.[592] Clearly, Provine was enormously successful in influencing his students to move toward atheism—an influence that made him very proud.

---

591 Provine, 1999, p. 123, emphasis added.
592 Provine, 1993, p. 63.

Creationists routinely run up against objections of 'separation of church and state' whenever they try to gain a hearing for their views in educational institutions. Provine's methods went beyond lecturing. He pressured students toward his atheistic views by assigning his own writings as required reading material, by meeting with them after class, and by other methods of intimidation and indoctrination. Yet the President of Cornell specifically forbids professors from presenting any evidence in favor of Intelligent Design, permitting only evidence in support of atheistic evolution in the university science classes. A reporter described Cornell's 2005 Presidential lecture under the headline "Cornell President Condemns Teaching Intelligent Design as Science"[593] as follows:

> A national movement to have Intelligent Design taught in science classrooms is "very dangerous," Cornell University's interim president, Hunter R. Rawlings III, said after taking up the issue Friday in a speech. But Mr. Rawlings charged that colleges were not engaging enough in the debate.
>
> Mr. Rawlings spoke to hundreds of faculty members and trustees during his state of the university address, which typically focuses on the college's accomplishments and business. Intelligent Design is a theory that says the universe is too complex to be the result of evolution and natural selection, proposing that a higher power is responsible.[594] Proponents say that alternatives to evolution should be taught in classrooms. But Mr. Rawlings denounced Intelligent Design as a "religious belief masquerading as a secular idea."
>
> Mr. Rawlings added, "Right now, this issue is playing out in school districts, cities, counties and states across the country." In citing a recent report by the Pew Research Center in Washington, Mr. Rawlings said 42 percent of Americans believe that creationism should be taught instead of evolution. "This is above all a cultural issue, not a scientific one," Mr. Rawlings said.
>
> "The Origin of Species" in 1859, … [and] the debate had resurfaced because the country is politically and culturally polarized. Cornell's staff should speak out [against] on any blurring of the

---

593 York, 2005.
594 This is an oft-quoted but misleading definition. ID is a positive inference from the evidence and from known causes, not a 'god-of-the-gaps' escape.

lines between religion and science.

John G. West, a senior fellow at the Discovery Institute in Seattle, which is a leader in the Intelligent Design movement, said he was concerned that Cornell's president was "fanning the flames of intolerance. A college president is in a unique position to create an atmosphere of free speech. If he's implying that faculty don't have the right to discuss ideas, I'm very concerned."

West was correct: by refusing alternative views of the scientific evidence, Rawlings has ensured that Cornell students will be openly indoctrinated only by Darwinism.[595] With such leadership at the helm of Cornell, it is little wonder then that Provine was never taken to task for his reckless and clearly illegal behavior.

References

Provine, William. 1993. "Response to Johnson Review." *Creation/Evolution*, Issue No. 32, Summer, 1993, pp. 62-63.

_____. 1999. "No Free Will." *Isis*. Supplement 117-123.

York, Michelle. 2005. "Cornell President Condemns Teaching Intelligent Design as Science." *New York Times*, October 22.
https://www.nytimes.com/2005/10/22/nyregion/cornell-president-condemns-teaching-intelligent-design-as-science.html

Videos

Will Provine comments – excerpt from the movie "Expelled." (6:07)
https://www.youtube.com/watch?v=MpJ5dHtmNtU

Darwinism: Science or Naturalistic Philosophy? (1:46:32)
Philip E. Johnson vs William Provine
Stanford University – April 30, 1994
https://www.youtube.com/watch?v=m7dG9U1vQ_U

---

[595] In all fairness, Provine did allow for some balance. He was friends with Phillip Johnson and sometimes had Johnson present his views in class. He also engaged in a famous debate with Johnson. His efforts to include creation and ID supporters extended to the likes of publisher and former Access Research Network blogger Kevin Wirth on his views for presentation to one of his classes. Some ID leaders respected Provine's openness to debate, wishing other Darwinists would follow his example. See these articles: https://www.evolutionnews.org/?s=Provine

Chapter 37

Chet Raymo, Ph.D.

From Catholic to Atheist

The cases detailed in this book reveal a consistent pattern: acceptance of Darwinism leads to rejection of Christianity and also quite often to atheism. This was clearly the case for Professor Chet Raymo. His reasons for becoming an atheist shed much light on the relationship between Darwinism and atheism. A well-known and highly respected popular science writer who has written over a dozen best-selling science books, Chet Raymo is now professor emeritus of physics and astronomy at the Catholic Stone Hill College in Massachusetts.

Professor Raymo was reared in a very religious Roman Catholic home. He attended Catholic schools until he went to graduate school. When Raymo's father was dying of cancer, he dealt with his condition by relying first on God, then on medical science and, lastly, on his own resources.[596] Raymo wrote that he believed his guardian angel hovered reassuringly at his side until it skipped from his conscience during adolescence. When he studied science, specifically Darwinism, it vanished completely.[597]

Life Under Darwinism

Like many other converts to evolution, Raymo came to believe that Darwinism makes God irrelevant.[598] In his words, he knew "the primary revelation of the Creator is the creation," and once Raymo became convinced that evolution was the creator, God was no longer needed.[599] After Raymo accepted Darwinism, he also figured that, if Genesis is wrong, then the entire Bible is unreliable. Consequently, he rejected it as well. In Raymo's mind, Darwinism murdered God. He read about how geologists once "struggled to find a way to make the story of the fossils compatible with the story of the Scriptures," but he decided that they had failed.[600] Once Darwinism had destroyed Raymo's theism, he became an evangelical atheist, actively preaching

---

596 Raymo, 1997, p. 110.
597 Raymo, 1998, p. 265.
598 Raymo, 1997, p. 108.
599 Raymo, 2004, p. 122.
600 Raymo, 1997, p. 108.

his new beliefs to the world through his college teaching, his writings, and his life.

Raymo admits that evolution "is not warm and fuzzy" and can even be "capricious and sometimes cruel."[601] He stresses that we should put aside our "security blankets" and accept the "cold and clammy truths" that we descended from amoebic ancestors and don't live in any kind of a nurturing universe.[602] Evolution is "relentless, inscrutable, and ruthless," — a view Raymo confesses comes right from Darwin himself:

> Humans are animals, Darwin believed, and like all animals they are locked in a struggle for existence, which, left to itself, eliminates the weak. Twenty-six years after [Darwin's daughter] Annie's death, Dr. Robert Koch took the first photograph ever published of a bacterium, the tuberculosis pathogen, and so confirmed the germ theory of disease. As Charles had guessed, Annie had died so that another creature might live.[603]

Raymo now actively opposes "religious people" who, he claims, see the world in black and white only, are comforted by dogma, and seek simple, certain truths. Raymo describes all religious people "true believers," i.e., as gullible people of faith, unable to be persuaded by reason or logic. Actually, many Christians became convinced theists after closely studying the evidence for the truth claims of Christianity. And many of them believe it is the evolutionists who are persuaded by speculation and therefore rely on faith-based evolutionary claims. Raymo recounts how Darwin's own theory had caused Darwin himself to see only the cruelty in nature and not its glorious designs. His negative focus "caused him to doubt the existence of an all-powerful loving God" and consequently the "promise of an afterlife."[604]

Raymo contrasts "believers" with "skeptics," whom he defines as people who hold their beliefs tentatively, are tolerant of others, and are more interested in refining their own views than in proselytizing others. He adds that if a "skeptic" is a theist, he or she must wrestle with God in a continuing struggle to hold onto theism. For this reason, he says (perhaps autobiographically), theists are often plagued by doubts.[605] He makes the outlandish claim that 100 percent of the scientific evidence favors Darwinism, and zero percent

---

601 *Raymo, 1998, p. 144.*
602 Raymo, 1998, p. 144.
603 Raymo, 2006, p. 106.
604 Raymo, 2006, p. 105.
605 Raymo, 1998, p. 3.

favors creationism.⁶⁰⁶ And yet most of the examples he provides are clearly incorrect. Raymo is given to irresponsible statements, such as: "I know of not a single article in the vast body of international, peer-reviewed scientific literature offering evidence for" creation.

First of all, there do exist papers in peer-reviewed scientific journals suggesting or supporting a design or creation view. Secondly, is it any wonder that today we don't see very much of that since any paper supporting a creation view submitted to most peer-reviewed science journals today is nearly always automatically discarded outright? Raymo's criticism is akin to admonishing a criminal for failing to show up at the local police station on a regular basis where his wanted poster is on full display. Finally, Raymo is either unaware or conveniently dismisses the huge problems within the peer review process itself as reported by no less than arguably the most prestigious peer-reviewed science journal of them all: Nature.⁶⁰⁷

Many Darwinians I have encountered frequently dismiss ID or creation-related investigations as unworthy because their research isn't published in their peer reviewed journals. Peer review is relied on, as Raymo does, by many evolutionists for the alleged reliability of valid scientific research. But it turns out it's well known that such reliance is seriously problematic and is therefore undependable. This is not an unusual or freakish concern, but many other publications have also echoed this claim as well for quite some time. ⁶⁰⁸ ⁶⁰⁹ ⁶¹⁰ ⁶¹¹ ⁶¹²

> The problem is that today's peer review is a broken process. Too often, errors slip through, and they can go uncorrected for years. Even if they are eventually exposed, that's often long after other researchers or clinical trials have relied upon them.⁶¹³

Richard Horton offers this scalding comment about peer review when he says

> The mistake, of course, is to have thought that peer review was any more than a crude means of discovering the acceptability—not the

---

606 Raymo, 1998, pp. 124, 125, 156.
607 Csiszar, 2016.
608 Manion, 2015.
609 McCook, 2006.
610 Berquis,. 2012.
611 Smith, 2006.
612 Shipka, 1987.
613 Pubpeer, 2014.

validity—of a new finding...We portray peer review to the public as a quasi-sacred process that helps to make science our most objective truth teller. But we know that the system of peer review is biased, unjust, unaccountable, incomplete, easily fixed, often insulting, usually ignorant, occasionally foolish, and frequently wrong.[614]

Raymo's misplaced reliance on peer review and his failure to acknowledge this decades-long problem should give readers considerable concern regarding the veracity of his other assumptions and beliefs regarding scientific matters.

After admitting that humans are staggeringly complex electrochemical machines, Raymo dogmatically asserts that there "is no ghost in the machine, no soul that exists independently of the body, and therefore no self that will survive the body's disintegration."[615] How he knows this from empirical science with such confidence is not stated. Blanket generalities such as these litter his writings, revealing that he himself is not a skeptic, but a "true believer" in his own religion (of atheism) by his own definition.

Raymo is either unaware of the vast body of evidence against his position, or is unwilling to acknowledge that evidence. Based on my experience with dogmatic atheists, both of these are likely true. Raymo now spends much of his time writing books on why the creation worldview is wrong, and why only the Darwinian worldview is "scientific." Actually, what he propagates is scientism. He bemoans the situation that children learn the 'fact' of Darwinism at school, but the 'myth' of creation at home. This, he thinks, confuses them about the 'fact' of evolution.[616] Such language is geared to intimidate, not enlighten.

It is clear from his writings that Raymo is not a product of objective education, but rather is a victim of Darwinian indoctrination.[617] What else could explain why he often ignores obvious evidence or makes misstatements, such as claiming that "Darwin was not an ardent Skeptic, but neither was he a True Believer. Evolution was *forced* upon him by his meticulous examination of the evidence."[618] In fact, after Darwin lost his theistic beliefs, he developed his theory to help him become an intellectually fulfilled Skeptic.[619]

Some of Raymo's statements indicate an appalling lack of knowl-

---

614 Horton, 2000.
615 Raymo, 1998, p. 134.
616 Raymo, 1998, p. 266.
617 Raymo and Raymo, 2001.
618 Raymo, 1998, p. 139.
619 See Jerry Bergman. *The Dark Side of Darwin*. Green Forest, AR: New Leaf Press, 2011.

edge, such as his claim that "the teaching of evolutionary biology is under nationwide assault by fundamentalist Christians, led by the powerful Traditional Values Coalition, a group that represents thousands of conservative churches."[620] Although the Traditional Values Coalition opposes the dogmatic teaching of Darwinism, they support teaching evidence against evolution and supporting design, and correcting the errors in irresponsible assertions such as Raymo's. They also evaluate and rate other organizations, including *Answers in Genesis*, the *Institute for Creation Research*, and the *Discovery Institute*, to name only three.

An example of Raymo's simplistic rhetoric is the following effort to disparage the creationist's view that no viable empirically-based theory exists to explain how eyes could have evolved. What "evidence" does Raymo give to support his view that eyes evolved purely by natural selection, mutations, time, and chance? Here it is: the fact that he needs glasses to read. This, he argues, shows that imperfect eyes can function, therefore one can imagine poor eyes improving over time by natural selection.

Medical experts say, to the contrary, that a nonfunctioning eye that does not see is often worse than no eye. From this flimsy argument, Raymo leaps to the generalization that a primitive eyespot in the one-celled organism *Euglena* (or some other progenitor) could have led to the human eye. This, he says, shows how evolution can be supported scientifically, while creationism cannot. He is apparently unaware of the work of I. L. Cohen, who, in his 1984 book showed that the human eye is irreducibly complex and could not have evolved by random mutations.[621]

Raymo's "proofs" for evolution are not scientific, but rhetorical. He repeatedly makes personal attacks on Darwin Skeptics. He also uses Richard Dawkins' "argument from personal incredulity" against them. For example, when creationists say it's highly improbable that evolution could get "from the goo-to-you-by-way-of-the-zoo," Dawkins responds by stating it is foolish to say that something is impossible just because it "seems impossible." "Nature conforms to the limits of our imaginations," he asserts with enormous bravado.[622] Such an appeal to imagination is unworthy of science. This is not evidence or proof, but a debating tactic.

Of course, *some things that seem impossible actually are impossible.* In science, evidence is required to determine if something is possible, and

---

620 Raymo, 1998, p. 141.
621 Cohen, I.L., *Darwin Was Wrong--A Study in Probabilities*. Greenvale, N.Y.: New Research Publications, 1984.
622 Raymo, 1998, p. 150.

much more evidence must establish a claim as true. Ridiculously, Raymo even claims that we cannot rely on the evidence of our senses or our mind, at least if such evidence contradicts Darwinism.[623] How, then, can he trust his senses to believe in evolution? It's a self-refuting argument. Raymo must empirically demonstrate the steps of the evolution leading from imperfect eyespots to complex eyes. He needs to identify the mutations on which natural selection can operate. It is not enough to "imagine" a set of changes that "might" do the job. This is what author and publisher Kevin H. Wirth refers to as "Probability Theology," where probabilistic notions based on imagination are held to as sufficient evidence to establish an evolution-based notion as entirely reasonable, if not accurate.

> It's worth noting that the Darwinists practice what I like to call "probability theology." Their belief in evolution, based on so many admitted probabilities, is more reverently adhered to than the familiar refrains memorized by any devoted Church Choirboy. The extent of how evolution "probably" occurred is on full display in work like the survey of fossil vertebrates found in eminent paleontologist Robert Carroll's book titled *Vertebrate Paleontology and Evolution*. In his comprehensive and wide ranging review of vertebrate fossils we find a state of constant conundrum among fossil investigators from all over the world who express uncertainty and even their inability to imagine how one fossil critter could have evolved into any other critter.
>
> Carroll's work is literally riddled on nearly every other page with comments from various workers primarily showcasing nothing more than imagination, supposition, extrapolation, and massive amounts of conjecture in their attempts to explain possible evolutionary pathways for nearly every creature found in the fossil record. The use of such terms as "may have," "must have," "probably were," "may have resembled," "presumably relied," "probably evolved," "may have evolved," "apparently arose," and so on are constantly used to describe what these experts believe.
>
> And since belief is not evidence, and also isn't a very sound basis of rational scientific thinking (an accusation frequently hurled against creationists and ID advocates), such beliefs qualify as a theology of sorts. To many evolutionists, that which is likely or

---

623 Raymo, 1998, p. 150.

probable is the equivalent of sound thinking and is even frequently referred to as solid and reliable evidence. But in fact, such comments are about as reliable as an ice cream cone staying frozen while sitting in the sun for hours on a hot summer day.[624]

Science requires testable evidence. No one, of course, has ever done that in detail: evolutionists just offer glittering generalities out of their own imaginations.

Raymo writes that what "seemed unlikely to Darwin, and seems impossible to creationists, has been shown to be quite reasonable by high-speed computer modeling. Not only reasonable, but given the proven premises of random mutations and natural selection, virtually inevitable."[625] It's not for Raymo to tell his readers what is reasonable; that is the job of the evidence. Like a salesman, Raymo claims that results of evolutionary models swiftly and decisively demonstrate the truth of Darwinism. Actually, the results say much more about the intelligent design of the programmers, and their assumptions. Without their help coaxing mutations toward progress and rewarding those reaching for the goal, macroevolution is extremely improbable, if not impossible. No evidence of "the proven premises of random mutations"—an area I have been researching for many years—is forthcoming in Raymo's writings.

In spite of enormous efforts to model Darwinian evolution, no one has ever demonstrated that mutations are a valid source of significant amounts of new information. Conversely, it has been empirically documented that decades of studies on fruit flies and other creatures show that almost all mutations are either near neutral or harmful.[626] One cannot use intelligently-designed computer programs operating on intelligently-designed computers to mimic evolution to prove evolution, which relies on the exact opposite of design: random, blind chance coupled with natural selection. Furthermore, computers cannot prove any other ancient historical set of events.

Raymo repeatedly uses examples of *microevolution* (such as those well-documented in Weiner's 1994 book)[627] as evidence for *macroevolution*, when he knows that *microevolution* is not in dispute by creationists or anyone else. Raymo relies on Andrew Dickson White's book, *The History of the Warfare Between Science and Theology in Christendom*, claiming that "little

---

624 Wirth, 2019 (in press).
625 Raymo, 1998, p. 152.
626 Bergman, 2005.
627 Weiner, 1994.

has changed since."628 Much has been written by both creationists and non-creationists about the egregious errors and distortions in the "warfare hypothesis" between science and religion. For Raymo to rely on White's now largely discredited book is irresponsible.

Despite his flippant disregard for scholarship, Raymo's writings have been reviewed favorably in journals ranging from *Science* to *Publishers Weekly* to *Astronomy* and *Choice*—and even by a Catholic priest! Yet the many major flaws in his evidence and reasoning have been glossed over by Raymo's reviewers, which included famous Harvard paleontologist, Steven J. Gould.

Those who fancy themselves "skeptics" while labeling others "true believers" ought to use the principles of skepticism to evaluate their own beliefs. Only then can they stand back and properly evaluate their circumstances fairly. As Darwin wrote in the *Origin* about "a fair result can be obtained only by fully stating…. on both sides of each question."629

Chet Raymo's careless writings might actually do his opponents a favor. For example, he employs intelligent design to argue there is no intelligent design. He commits logical fallacies to argue that creationism is fallacious. He demands scientific evidence, but dishes out questionable generalities based on his own imagination. His writings, pregnant with quotations that support much of what Darwin critics have been saying for decades, can be extremely valuable for creationists looking for ammunition to show the tragic consequences of indoctrination into evolution. Raymo is Exhibit-A for showing how Darwinism is a doorway to atheism.

For example, Raymo carelessly asserts that "everything [everything?] science has learned since Galileo suggests that we are accidental, contingent, ephemeral parts of creation, rather than lords over it."630 This self-refuting claim (e.g., if Raymo is a product of accidents, are his beliefs also accidental?) expresses what Darwin critics have been stressing for decades–i.e., Darwinism impacts one's worldview and philosophy of life. Although Raymo claims not to be a critic of religion, his work belies the claim. He dogmatically states that we humans "are not immortal," but "fleeting," and that our "spirits are the brief efflorescence of complexity."631

One wonders how Raymo knows all of this from materialistic science. He observes that many educated people in the West, including himself, long for something akin to religious faith, but can neither accept the idea of God,

---

628 Raymo, 1998, p. 159.
629 Darwin, 1859, p. 2.
630 Raymo, 1998, p. 163.
631 Raymo, 1998, p. 245.

nor quite leave it alone. Raymo's religion is that of scientism. He even stated that photographs of the universe to him are "religious" icons that expand our horizon and sharpen our senses about the enormity and beauty of the universe.[632]

Raymo writes that, "skepticism is a critical reluctance to take anything as absolute truth," then dogmatically asserts that humans are the offspring of comets.[633] A page later, he says that "the heron like all birds is a close relative of dinosaurs, and that feathered birds first flapped their wings in Jurassic times," never hinting that bird evolution from dinosaurs is very controversial for many good reasons.[634]

Even a passing familiarity with paleontology produces an awareness of the great debate between eminent paleontologists over the origin of birds. Some "experts" argue that birds evolved from dinosaurs, while others argue that they evolved from non-dinosaurian reptiles. New discoveries throw even more confusion into the mix, indicating that the DNA of birds may be closer to that of mammals than to that of either reptiles or dinosaurs.[635]

Raymo claims that if there was solid evidence supporting the creation worldview, scientists "would be falling over each other to publish it.... Every scientist I know is as happy to have something proved wrong as proved right. Either outcome advances us toward truth."[636] Surely Raymo cannot be this naïve. In a perfect world this would be true, but historians of science can document many cases where new ideas were strenuously resisted, even when the research was very persuasive. Among the many examples include the discovery that *Helicobacter pylori* is the cause of the majority of ulcers—not stress or excess stomach acid as was once universally believed. In another case, the consensus fought J. Harlen Bretz for decades over his theory that the Channeled Scablands were produced by a megaflood. Semmelweis was ridiculed for advising doctors to wash their hands after leaving the morgue before delivering babies; Pasteur was ridiculed for opposing spontaneous generation. True, the correct views eventually prevailed, but not without enormous difficulty in spite of abundant medical and scientific evidence. How could one know this is not happening again with Darwinism?

---

632 Raymo, 1998, p. 243. Also see Bergman, 2017.
633 Raymo, 1998, p. 251.
634 Raymo, 1998, p. 254.
635 Bergman, 2017.
636 Raymo, 1998, p. 146.

References

Bergman, Jerry. 2017. *Fossil Forensics: Separating Fact from Fantasy in Paleontology.* Tulsa, Oklahoma: Bartlett Publishing

Berquist, Thomas H. 2012. "Peer Review: Is the Process Broken?" *American Journal of Roentgenology* (AJR), August 2012, 199(2);2. (Editorial)

Cohen, I.L., 1984. *Darwin Was Wrong--A Study in Probabilities.* Greenvale, N.Y.: New Research Publications.

Csiszar, Alex. 2016. "Peer Review: Troubled from the Start." *Nature*, 532(7599): 306-308. April 19.

Darwin, Charles. 1859. *The Origin of Species.* London: John Murray.

Horton, Richard. 2000. "Genetically Modified Food Consternation, Confusion, and Crack-up." *The Medical Journal of Australia.* 172(4).

Manion, Kieran. 2015 "Peer Review Is Broken." *IMMpress Magazine*, September 27. http://www.immpressmagazine.com/peer-review-is-broken/

McCook, Alison. 2006. "Is Peer-Review Broken?" *The Scientist*, February 1. https://www.the-scientist.com/?articles.view/articleNo/23672/title/Is-Peer-Review-Broken-/

Raymo, Chet. 1997. *Honey from Stone.* Kerry, Ireland: Brandon.

_____. 1998. *Skeptics and True Believers.* New York: Walker.

_____. 2004. *Climbing Brandon: Science and Faith on Ireland's Holy Mountain.* New York: Walker.

_____. 2006. *Walking Zero: Discovering Space and Time Along the Prime Meridian.* New York: Walker.

_____ and Maureen E. Raymo. 2001. *Written in Stone: A Geologic History of the Northeastern United States.* New York: Black Dome Press

Shipka, Tomas. 1987. "The Crisis in Peer Review," *NEA Advocate*, Apr.-May, pp. 6-7.

Smith, Richard. 2006. "Peer review: a flawed process at the heart of science and journals," *Journal of the Royal Society of Medicine.* 2006 April, 99(4):178–182.

Weiner, Jonathan. 1994. *The Beak of the Finch: A Story of Evolution in Our*

Time. New York: Knopf.

Wirth, Kevin (editor). 2019. *Slaughter of the Dissidents: Compendium.* Leafcutter Press (in press)

Chapter 38

Stanley Arthur Rice, Ph.D.

A former Creationist, now an Atheist

Rice is the author of *Encyclopedia of Evolution*[637] and many other scientific publications. He wrote that he was a creationist when a biology major at the University of California at Santa Barbara. Why he left both Christianity and theism, he explains, as follows:

> Many of you have probably had the same experiences I had. As I learned about evolution, I had to give up creationism. As I read the Bible and recognized its human authorship, I had to give up simple biblical faith. But it was not easy. I wanted to cling to something. The reports of near-death experiences gave me something I thought I could cling to. Were these people really seeing into Heaven? If so, then there really was a Heaven. I did not need to base such a belief on questionable biblical passages; I could base it on the eye-witness accounts of people alive today. But how do you know they were not hallucinating? Oxygen deprivation could make their brains create all these sensations.[638]

He also claims that it was the overwhelming evidence for evolution that converted him first to theistic evolution, then to atheistic evolution. Rice also now rejects Intelligent Design, derisively calling it pseudoscience.[639] To illustrate his complete apostasy from Christianity, Rice stated his opinion that religion can have a very negative influence on science because "it can disrupt the ability to reason from evidence." As an example of what he means, he claims that:

> Creationists have long used, and still like to use, the "sudden appearances in the fossil record" argument. Botanist George Howe … addressed a small crowd in a big auditorium at the University of California, Santa Barbara in 1976. A young, impressionable creationist student named Stan Rice was in the audience. Howe's main point was that angiosperms suddenly appeared, without rec-

---

637 Rice 2007.
638 Rice, 2011, p. 168.
639 Rice, 2011, pp. xv, 213-217.

ognizable ancestors, during the Cretaceous Period.[640]

Rice says that there exist two major problems with this creationist argument, namely

> most botanists would say that, indeed, there are recognizable ancestors,[including the] the Bennettitalean conifers. ...The resemblance was not close enough to convince Howe. ...These obvious flaws in creationist reasoning were invisible to the people present at the talk—at least to me. My religious zeal blinded me to these flaws, which I recognized only years later.[641]

Howe's claim, Rice decided, was false. So, he became an evolutionist. Unfortunately, he didn't look further into the fossil record where he would have learned that most paleontologists can only speculate about the evolutionary history of nearly all vertebrates, since the fossils clearly show stasis and stability rather than evolution.[642] He opined that a creationist can look directly at the evidence for evolution and interpret it, not as evidence for evolution, but instead as evidence for the Genesis Flood. He complains that

> fundamentalist students in college classes (who are generally very nice people) can look right at the evidence and choose to not believe it. And some of them are very smart. The 2012 valedictorian at Southeastern got the best grade in my evolution class, but remained a creationist.[643]

---

640 Rice, 2014, p. 10.
641 Rice, 2014, p. 20.
642 The writers of most books on surveys of vertebrate evolution spend the vast majority of their time providing detailed descriptions of fossil organisms and very little time (if any) describing what can only be described as "presumed" evolutionary relationships. Why? Because they are in a constant state of disagreement regarding who the precursors are for every major animal found. Why? They blame an 'incomplete' fossil record, but they never stop to consider that they are looking for missing organisms they will never find, because the intermediates they expect simply never existed. Time and time again they are forced to admit they have no clue about who evolved into what. Nearly all of them admit that they are baffled by their inability to piece together evolutionary relationships with comments like these: "Echinoderms have perhaps the most distinctive body plan of any animal group – so distinctive, in fact, that their abundant fossil record doesn't shed as much light as one would think on their origins." Actually, the fossil record doesn't shed any light on their origins. This author simply isn't willing to admit it. Gee, Henry. 2018. *Across the Bridge*. Chicago: University of Chicago Press.
"The fossil record does not provide evidence of the transition toward either pterosaurs or bats: the earliest known members of these groups had already evolved an advanced flight apparatus." This type of remark is common when describing the ancestry of nearly all known organisms found in the fossil record. – K. Wirth, editor
Carroll, Robert. 1997. Patterns and Processes of Vertebrate Evolution. Cambridge University Press.
643 Rice, 2014, p. 21.

According to his writings, Rice is now a militant atheist who sounds a lot like Richard Dawkins in his condemnation of Christianity. Some of his anti-creation and other publications are as follows:

"Botanical and ecological objections to a pre-flood vapor canopy," *Journal of the American Scientific Affiliation*, Vol.37 (1985), pp. 225-229.

"Creationist ecology?" *Bulletin, Ecological Society of America*, Vol.67 (1986), pp. 8-10.

"On the problem of apparent evil in the natural world," *Perspectives on Science and Christian Faith*, Vol.39 (1987), pp. 150-157.

"Scientific creationism: Adding imagination to scripture," *Creation/Evolution* XXIV (1988), pp. 25-36.

"Bringing goodness out of adversity: God's activity in the world of nature," *Perspectives on Science and Christian Faith* 41 (March 1989), pp. 2-9.

"Faithful in the little things: Creationists and 'operations science'" *Creation/Evolution* XXV (1989), pp. 8-14.

References

Stanley Rice. 2007. *Encyclopedia of Evolution* New York: Facts on File, 2007

_____. 2011. *Life of Earth: Portrait of a Beautiful, Middle-Aged, Stressed-Out World*. Amherst, NY: Prometheus Books.

_____. 2014. "Confessions of an Oklahoma Evolutionist: The Bad, the Ugly and the Good." *Reports of the National Center for Science Education*. 34(1):18-24. January February.

Chapter 39

Alan Rogers, Ph.D

Former Creationist Now a Vigorous Creation Opponent

Dr. Alan Rogers was born in Texas where his father was a Southern Baptist minister. Rogers' father eventually left the ministry to practice clinical psychology, the field in which his mother also worked. When the family moved to Charleston, West Virginia, they began attending the more liberal American Baptist church.

Young Rogers went on to become an anthropologist, specializing in population genetics and evolutionary ecology. He received an undergraduate degree from the University of Texas-Austin, and a doctorate from the University of New Mexico. His college education convinced him that his creation worldview was scientifically wrong. During his studies, he learned that his creationist uncle had made the same argument that 17th-century naturalist Philip Henry Gosse advocated in the mid-1800s, namely that the *appearance of design* theory satisfactorily explained the long ages postulated by Darwinists and others—and Rogers felt that this idea had been debunked in the decades since Gosse, the most well-known advocate of this theory.[644]

Dr. Rogers began his teaching career in 1988, when he was hired to teach biology at the University of Utah. It wasn't until around 2006 that Rogers, after reading a poll that reported only about half of Americans believe humans evolved, felt the need to share the "proof" for human evolution that he learned in his undergraduate *Evolution of Human Nature* course where he was converted to Darwinism. Rogers then began spending a week or two in his introductory classes focusing on what he alleged was the evidence for evolution.

Unable to find a textbook that he felt was adequate to help him reach this goal, he authored one, a short book titled *The Evidence for Evolution*, published in June of 2011 by the University of Chicago Press. His book was endorsed by none other than Steven Pinker of Harvard University who stated "anyone with an open mind would find this book convincing." This text covers the usual refuted, "evidences" for evolution usually trotted out in most high school biology textbooks, including the backward retina, the poorly-designed vas deferens, whale evolution, the genetic clock, and the fossil record. It is a

---

[644] This claim is not accurate. The appearance of design view is widely accepted by many creationists and others. See http://www.creationism.org/heinze/appearanceofdesign.htm

pity that evolutionists like Rogers are unaware that these evidences strongly support design, not Darwinism.

References

Skolnick, Deena, et al. 2018. "No Missing Link: Knowledge Predicts Acceptance of Evolution in the United States." *BioScience*, 68(3):212-222, March 1.

Video: The Evidence for Evolution (45:25)
https://www.youtube.com/watch?v=4-bZBqy_VZkv

Chapter 40

Nicolaas Adrianus Rupke, Ph.D.

Creationist to Evolutionist

In a book on the history of creationism in Europe, Rupke writes that for a number of years he was involved in the development of creationism theory. When in 1961, as an aspiring young-Earth creationist but still in high school, he was exposed to creationism in the writings of George McCready Price and soon corresponded with him. He later learned about the work of Dr. Henry Morris and soon carefully read the *The Genesis Flood*, co-authored by Henry Morris and Princeton graduate, Dr. John Whitcomb. As an undergraduate student, he studied the history of biblical catastrophism by reading the writings of various evolutionary and other scholars, including John Ray, John Woodward, Jean Andre Deluc, and even George Fairholme. He later was reintroduced to some new examples for

> the old argument for cataclysmal sedimentation in the form of polystrate dendrolites. At the same time, I enthusiastically took part in mainstream neo-catastrophist sedimentology, just then gaining ground in such forms as the turbidity current theory put forward by my mentor Philip Henry Kuenen. In many ways, neo-catastrophist science seemed to reconfirm a Judeo-Christian picture of the history of the world. It has given succor and support to creationism—young-earth, old-earth, also the more recent Intelligent Design movement.[645]

Articles that he published in creation magazines include "Prolegomena to a Study of Cataclysmal Sedimentation," *Creation Research Society Annual 3*;[646] reprinted in Walter E. Lammerts, ed., *Why Not Creation?*[647] Secular geology and the question of the age of the Earth were important in his conversion. He later became convinced that evolution was fully confirmed by science and then "left the creationist movement," never to return.[648] No doubt part of

---

645 Rupke, 2014, p. 248.
646 Rupke, "Prolegomena to a Study of Cataclysmal Sedimentation," *Creation Research Society Annual* 3,1966, pp. 16-37.
647 Lammerts, pp. 141-179.
648 Rupke, 2014, p. 249.

the reason why, was because he completed his Ph.D. in the Department of Geological and Geophysical Sciences at Princeton University which stresses Darwinism. I could find no evidence one way or the other relating to his theological beliefs. Some of his many publications include:

1. "Late Quaternary rates of abyssal mud deposition in the western Mediterranean Sea" *Marine Geology* 17: M9-16 (1974) with D.J. Stanley & R. Stuckenrath.

2. Rupke, N.A. "Prolegomena to a Study of Cataclysmal Sedimentation," CRSQ, Vol.3 (1996).

3. "A summary of the Monera fallacy" CRSQ 4(3): 106-113 (1967) reprinted in 'Scientific Studies in Special Creation' (1971), St. Joseph, MO: CRS Books, pp. 169-183.

4. "Sedimentary evidence for the allochthonous origin of stigmaria, Carboniferous, Nova Scotia" *Bulletin of the Geological Society of America* (1970), 81(8): 2535-2538.

5. "Continental drift before 1900," (25 July 1970), *Nature*, 227:349-350.

6. "Distinctive properties of turbidic and hemipelagic mud layers in the Alge'ro Balearic Basin, western Mediterranean," *Smithsonian Center of Earth Sciences* (1974), 13:40 with D.J. Stanley.

7. *Vivisection in Historical Perspective* (1987), Editor.

8. *Richard Owen: Victorian Naturalist.* (1994), New Haven, CT: Yale University Press.

9. "Richard Owen's vertebrate archetype," (1993), *Isis*:84

10. *The Great Chains of History: William Buckland and the English School of Geology*, 1814-1849, (1983), Oxford: Clarendon Press.

11. Article in *Journal of Geoscience Education*, Vol. 32.

12. "Eurocentric ideology of continental drift" *History of Science* 34:1-22 (1996).

13. "Deposition of fine-grained sediments in the abyssal environment of the Algero-Balearic Basin, western Mediterranean Sea," (1975), *Sedimentology*: Vol. 22.

14. " *Bathybius haeckelii* and the psychology of scientific discovery. Theory in-

stead of observed data controlled the late 19th century 'discovery' of a "primitive form of life," *Studies in History and the Philosophy of Science*, 7(1).

15. *Science, Politics and the Public Good* (London, 1988), Editor.

16. "The study of fossils in the romantic philosophy of history and nature" *History of Science*, (1983), Vol. 21, pp. 389-413.

17. "The apocalyptic denominator in English culture in the early nineteenth century," in M.R. Pollock, Editor, *Common Denominators in Art and Science* (1983) Aberdeen University Press, pp. 30-45.

18. "Romanticism in the Netherlands," in R. S. Porter and M. Teich, Editors, *Romanticism in National Context* (1988), Cambridge: Cambridge University Press, pp. 191-216.

19. Vitae - https://www.wlu.edu/Documents/directory/rupken.pdf?v= 635954397342299151

20. Video: "Huxley's Rule" and the origins of scientific racism. 2015. (31:58) https://www.youtube.com/watch?v=koFbc1RD5Go

References

Bergman, Jerry. 2012. Hitler and the Nazi Darwinian Worldview: How the Nazi Eugenic Crusade for a Superior Race Caused the Greatest Holocaust in World History. 2012. Kitchener, Ontario, Canada: Joshua Press.

Forrest, Barbara and Paul Gross. 2004 Creationism's Trojan Horse. New York: Oxford University Press, pp. 103–104.

McCabe, Joseph and Thaddeus Burr Wakeman. 1911. The Answer of Ernst Haeckel to the Falsehoods of the Jesuits. New York: The Truth Seeker Company.

Lammerts, Walter E., ed. 1970. Why Not Creation? Philadelphia: PA. Presbyterian and Reformed Publishing.

Rupke, Nicolaas. 2014.

in Stefaan Blancke and Hans Henrik Hjermitslev. 2014. Creationism in Europe (Medicine, Science, and Religion in Historical Context). Baltimore, MD: Johns Hopkins University Press. Rupke appears in chapter titled "Afterword" pp. 142-149.

Spiro, Jonathan. 2009. Defending the Master Race. Burlington, VT: University of Vermont Press

Chapter 41

Michael Shermer, Ph.D.

From Ministerial Student to Atheist

Dr. Michael Shermer was introduced to Christianity as a youth, and in his senior year of high school he professed faith in Christ as his Savior at the behest of a close friend.[649] The next day, he attended church and made a public profession of his new faith. He then studiously read the Bible and books about the Bible, regularly attended youth church groups (such as The Barn in La Crescenta, California), and gathered with fellow Christians a couple of times a week to sing Christian songs and worship. Headed for the ministry, he enrolled in Pepperdine University (a Church of Christ school) to major in theology. Although he went to chapel and prayed regularly, as a new Christian he still had some questions—such as the problem of evil.

While attending Glendale College, the budding minister took a philosophy class from Professor Richard Hardison. Deciding to witness to his professor, Michael gave him a book on Christian theology. The professor took it upon himself to refute the book, and typed out a list of problems with the book's theme that he gave to Michael. Soon followed many long discussions, both in and after class, in which Hardison eventually won Michael over to atheism. He thus converted from evangelical Christianity to evangelical atheism, and is now active in proselytizing against both Christianity and theism.[650] He writes that in the late 1970s he

> read the Bible very carefully as a theology student at Pepperdine University (before I switched to psychology), and like many in the early 1970s, I had been a born-again Christian, taking up the cause with considerable enthusiasm, including "witnessing" to non-believers.[651]

Then, during his graduate course work in

> ethnology (the study of animal behavior) at California State Uni-

---
649 Shermer, 2000, *How We Believe: The Search for God in an Age of Science*. NY: Freeman,. p. 2.
650 Shermer, 2000, pp. 6-9.
651 Shermer, 1997. Chapter 9 "In the Beginning: An Evening with Duane T. Gish," p. 127.

versity, Fullerton, I ran into the brilliant ... Bayard Brattstrom and the insightful and wise Meg White. Brattstrom was far more than one of the world's leading experts in behavioral herpetology (the study of reptilian behavior). He was well versed in the philosophical debates of modern biology and science, and regularly regaled us for hours with philosophical musings over beer and wine at the 301 Club (named for the nightclub's address) after the Tuesday night class.[652]

The result was:

> Somewhere between Brattstrom's 301 Club discussions of God and evolution and White's ethological explanations about the evolution of animal behavior, my Christian icthus [sic (ichthys}] (the fish with Greek symbols that Christians wore in the 1970s to publicly indicate their faith) got away, and with it my religion. *Science became my belief system, and evolution my doctrine.*[653]

Shermer now strenuously opposes all attempts by believers to "use science and reason to *prove* God's existence."[654] Ironically, as Editor of *Skeptic Magazine* and author of numerous books, he spends a great deal of time using "science and reason" to *disprove* (or at least to argue against) God's existence. He is especially active in attacking creationism because, in his words,

> the number-one reason people give for why they believe in God is ... the classic cosmological or design argument: The good design, natural beauty, perfection, and complexity of the world or universe compels us to think that it could not have come about without an intelligent designer. In other words, people say they believe in God because the evidence of their senses tells them so.[655]

Because of the design argument's power, many professors (like Hardison) attack the classic cosmological argument in order to win students over to atheism. Michael openly acknowledges that he was set along this path by his philosophy professor, whom he calls one of his best professors (and, he adds, is "now my friend").

---

652 Shermer, 1997, pp. 127-128.
653 Shermer, 1997, pp. 127-128; emphasis added.
654 Shermer, 2000, p. xiii.
655 Shermer, 2000, p. xiv.

Although Hardison has also been very active, and very successful as well in converting students to his atheistic worldview, I could not find any record of complaints or concerns about his overt proselytizing activity. Indeed, he is regarded as an excellent teacher who is sincerely interested in his students, even as he openly challenges their Christian faith and evidently not infrequently, wins students over to his views. It is possible that some students have objected, feeling that he is proselytizing against religion, but their complaints were not on record and have never made it to court (and, if they did, the ACLU and the other organizations would likely vigorously defend Hardison's academic freedom). He has now spent most of his life as an evolutionary apologist.

References

Shermer, Michael. 1997. *Why People Believe Weird Things: Pseudoscience, Superstition, and Other Confusions of Our Time.* New York: Freeman.

_____. 2000. *How We Believe: The Search for God in an Age of Science.* New York: Freeman.

Videos

Controversies in Science (2012) with Michael Shermer (56:46)
https://www.youtube.com/watch?v=4-Krxa1q00w

The Believing Brain: How we construct beliefs and reinforce them as truths (1:08:46)
https://www.youtube.com/watch?v=R6ijdDtOLLo

Debate: *What Best Explains Reality: Theism or Atheism?* (Frank Turek vs. Michael Shermer) (2:04:08)
https://www.youtube.com/watch?v=8aZn7XUFSmA

Ben Shapiro Show (1:01:59)
https://www.youtube.com/watch?v=ZaxUG3n1KMA

God Does Not Exist (16:03)
https://www.youtube.com/watch?v=0pOI2YvVuuE
Where he states that Humans created God, not vice versa.

Chapter 42

George Gaylord Simpson, Ph.D.

From Creationist to Leading Evolutionist

George Gaylord Simpson, was the preeminent paleontologist and evolutionary biologist for over two decades after the end of World War II. He was one of the most influential paleontologists of the twentieth century, and a major participant in achieving the modern evolutionary synthesis, often called Neo-Darwinism, that replaced the old Darwinism. His major books include *Tempo and Mode in Evolution* (1944), *The Meaning of Evolution* (1949) and *The Major Features of Evolution* (1953), all of which have in the past been required reading at many colleges.

Although Simpson was reared a fundamentalist Christian, by his early teens he had rejected Christianity, and eventually became an agnostic.[656] As a youth, he regularly went to triple Sunday church services plus the midweek prayer meetings. He writes that his family was religiously more active than most, commonly attending

> three services on Sundays as well as the midweek prayer meeting, and in addition had family prayers and psalm recitations. At the age of nine I formally joined the church, having assured the minister that I believed the Presbyterian creed.[657]

He met his future wife, Ann Roe, when they both were still very young, but the relationship cooled somewhat when she briefly flirted with Christianity due to the influence of friends. As she learned about Darwinism in college and from friends, she rejected Christianity, and they later married. Ann Rose has since written a book documenting her finding that most all leading scientists are atheists, or at least agnostics.[658] Her data came from extensive interviews with and the testimony of eminent scientists.

The reasons Professor Simpson left Christianity and theism include his observation that he believed a person can be fairly

---

656 Simpson, 1987, p. 16.
657 Simpson, 1987, p. 25.
658 Roe, 1952.

certain that a belief contrary to the weight of factual evidence is not true. ...there are indeed many known facts bearing on the origin of man, and these facts are strongly inconsistent with the creation of man just so, in his present form, and hence in the image of any given being or Being. ...The long magnificent story of man's evolution from the stuff of stars is incomparably more wonderful, more awe-inspiring than that fable of an anthropomorph making mudpies in a mythical Garden of Eden.[659]

Having knocked down a straw man with ridicule, and embraced scientism's glittering generalities, he decided that "far from being able to explain all known material phenomena ... we cannot rationally hope to proceed toward explaining them in any way but by naturalistic interpretation of observed facts."[660] Thus, he abandoned his childhood faith for the worldview that only naturalism can explain life, and that no evidence exists to support the notion of an intelligent creator. He never quite gets around to explaining how rationality evolved.

References

Bergman, Jerry. 2016. *Debunking Human Evolution Taught in Our Public Schools* with Dr. Daniel A. Biddle and David A. Bisbee. Genesis Apologetics. Folsom, CA. 2nd edition.

Roe, Ann. 1952. *The Making of a Scientist*, New York: Dodd, Mead and Company.

Simpson, George Gaylord. 1964. *This View of Life: The World of an Evolutionist*. New York: Harcourt, Brace and World, Inc.

_____. 1978. *Concession to the Improbable: An Unconventional Autobiography*. New Haven and London: Yale University Press.

_____. 1987. *Simple Curiosity. Letters from George Gaylord Simpson to his Family 1921-1970*. Berkeley, CA: University of California Press.

---

659 Simpson, 1987, p. 29. Here he ignores the many problems with the ape to human theory. See Jerry Bergman, 2016. Chapter 4, *What about the Different Races of People*, pp. 83-92; Chapter 5 with Jeffrey Tomkins, *Are Humans and Chimps 98% Similar?* Pp. 93—111; chapter 7 *The Problem of Overdesign for Darwinism* pp. 116-124; and chapter 8 *Vestigial Structures in Humans and Animals* pp. 125-138.
660 Simpson, 1987, p. 29.

Chapter 43

Howard N. Teeple

From Conservative Christian to Evolutionist

Howard N. Teeple was reared as a Christian fundamentalist, and was very involved in his church for most of his youth. Soon after he professed to receive Christ, he headed off to Willamette University in Oregon where a professor taught the theory of evolution as fact. It didn't take long for Teeple to reject his Christianity. It began as a result of taking a Bible history course taught by Herman Clark, an ordained Methodist minister with a Ph.D. in religion from the University of Chicago. The course "shook me up psychologically" and "opened" his eyes to the "fallacies of fundamentalism," thereby changing the direction of his life forever.[661]

According to Teeple, Professor Clark worked hard at demolishing his students' trust in the Scriptures, especially in his course titled "Records of a Life of Jesus." This experience had a major influence in Teeple's loss of faith. Teeple noted that "by the end of the course the students, including myself, generally accepted the liberal point of view, although we differed among ourselves in our selection of the Gospel tradition of what is authentic, that is, of what is an accurate report of what Jesus said and did."

Given the intellectually-dishonest option by Professor Clark to allow students to decide for themselves if they wanted to accept or reject portions of the Bible, the students arbitrarily picked what they liked or rejected, for whatever reason they fancied—somewhat like obeying the laws one agrees with, and ignoring those one dislikes or disagrees with. No other classic book gets this kind of revisionist treatment; otherwise, intellectual history would break down entirely. (Some would say that's exactly what happened in the 'post-modernism' fad of the 1980s.)

Teeple taught at a number of theological schools, eventually retiring as a full professor in 1977, from Chicago State University. Undoubtedly, his tenure influenced many of his students to accept Darwinism. His books include *The Noah's Ark Nonsense*, which used similar arguments employed in a book by a colleague of his, Dr. Lloyd R. Bailey of Duke University, who wrote *Where's Noah's Ark?* Ironically, St. Peter prophesied that scoffers would come

---

661 Teeple, 1995, pp. 348–349.

in the last days, denying Creation and the Flood (II Peter 3:3-6). Now we see why; Darwinian evolution has been a major factor in undermining faith in the Scriptures and even in the existence of a Creator. Unbelief and apostasy are nothing new. What's new is Darwinism giving a pseudo-scientific justification for unbelief.

Reference

Teeple, Howard. 1995. Chapter 6 page 347-359 in Edward T. Babinski's 1995. *Leaving the Fold: Testimonies of Former Fundamentalists*. Amherst, NY: Prometheus Books, quote from p. 349.

Chapter 44

Charles Templeton

Preacher becomes an Atheist by way of Darwinism

Former evangelist and Billy Graham co-worker Charles Templeton was one of America's leading evangelists. In 1945, Templeton and Torrey Johnson of Chicago, Illinois met with a number of youth leaders from around the United States at Grace College in Winona Lake, Indiana. Their goal was to form an organization, which is now known as Youth for Christ, which was formed in 1946. Torrey Johnson was elected as its first President, and Billy Graham was the first full-time evangelist.

Charles Templeton's crusades were attended by tens of thousands of people. His last sermon, given in the fall of 1954 at Harrisburg, PA, drew one of the greatest crowds ever to gather in the history of Harrisburg. The reason he left the ministry was not due to his lack of success, but rather because he came to believe that it was beyond dispute that "all life is the result of timeless evolutionary forces," and not a Creator. A major reason why Templeton became an atheist was due to his acceptance of Darwinism, specifically because he came to believe that:

1. There is no God in the biblical sense—rather all life is the result of timeless evolutionary forces, having reached its present transient state over millions of years.

2. The genus *Homo* is only the leading edge of the universal evolutionary process, but has more capacity for good and evil than any other creature.

3. There exists what he described as a Life Force, ... a Primal Energy, or Life Essence, which is the genesis of all that exists, from the simplest atom to the entirety of the expanding universe. This Life Force is not a person or even a "being." It does not love nor can it be loved, it simply exists. Where this Life Force came from he never says.

4. It is likely that there are millions of populated worlds in the universe, each at different stages of evolution, some dying, some being

born. Our world was born in finite time and, in time, will perish.[662]

How he arrived at most of these conclusions was not identified in his writings, but it is clear that Darwinism was critical to many, if not most of them. Lacking formal theological education, he decided to pursue a theology degree at Princeton Theological Seminary. Resigning from the church he had pastored for several years, Templeton began his coursework at Princeton in 1948 where he was exposed to evolution and doubts about Genesis. After he graduated, he accepted a position with the very liberal National Council of Churches and eventually left the ministry altogether. No doubt reading general popular literature on Darwinism was also integral in his conversion to Evolution. As a journalist, novelist and prolific writer, Templeton read widely.

Lee Strobel conducted an in-depth interview with him shortly before he died.[663] In that interview, Strobel related that Templeton had some of the same concerns that Strobel had, which included Darwinism, the lack of evidence for Christianity, and the problem of evil. The difference is, Strobel went on to resolve these issues, whereas Templeton did not. Even when dying, Templeton refused to reconsider his rejection of God, openly missing out on the comfort that his faith once gave him.[664]

References

Strobel, Lee. 2000. *The Case for Faith*. Grand Rapids, MI: HarperCollins.

Templeton, Charles. 1996. *Farewell to God, My Reasons for Rejecting the Christian Faith*. Toronto: McClelland and Stewart.

---

662 Templeton, 1996, p. 232.
663 Strobel, 2000.
664 Strobel, 2000, pp. 18, 22.

Chapter 45

Howard Van Till, Ph.D.

From Conservative Christian to Griffin's Naturalism

Howard J. Van Till is Professor Emeritus of Physics and Astronomy at Calvin College, Grand Rapids, Michigan. After graduating from Calvin College in 1960, he earned his Ph.D. in physics from Michigan State University in 1965. Van Till was reared as a conservative Christian and has been active in writing and speaking on the interaction of science and Christian belief for decades. His 1986 book, *The Fourth Day*, was an attempt to deal with the creation-evolution controversy by rejecting both creationism and evolutionism.[665] About this time he was a member of the editorial boards of the now liberal *Science and Christian Belief*[666] and *Theology and Science*,[667] and also worked with the Templeton Foundation. No doubt this involvement was important in his journey from theism to naturalism.

Author T. E. Wilder wrote in 1991 that in fact, Van Till challenged "the Biblical basis for rejecting the theories of cosmic and human origins now dominant in science."[668] On October 23, 2004, Howard Van Till presented a talk titled *From Calvinism to Claremont: Now That's Evolution! One Scientist's Evolution from Calvin's Supernaturalism to Griffin's Naturalism*.[669] On May 24, 2006, he gave a presentation to the Freethought Association of West Michigan titled *From Calvinism to Freethought: The Road Less Traveled* in which he recounted that on several key Christian questions, he now was far less certain than before. In short, he believed that we should accept the current orthodox science worldview, and reject the traditional Christian view. He writes that he knew some members of the Calvin College community

> might object to my friendly attitude toward the word 'evolution' and to my judgment that Calvinist theology left more than enough

---

665 Van Till. 1986. p. 217.
666 Science and Christian Belief is a Journal that is "concerned with the interactions of science and religion, with particular reference to Christianity." https://www.scienceandchristianbelief.org/about.php
667 Theology and Science "engages scientific discourse in dialogue with both Christian and multi-religious perspectives. With these affiliations, the journal provides a critical and comprehensive collection of articles and reviews that promote the creative mutual interaction between the natural sciences and theology." http://www.ctns.org/theology_science.html
668 Wilder, 1991, p. 37.
669 Held at the Claremont School of Theology, Claremont, California, 9/17/2011.

> room for it. Very naively, however, I thought that these objections would be expressed politely as carefully-crafted intellectual arguments that I could fruitfully engage in the familiar style of the academy. I was dead wrong.

As a result of his new evolutionary ideas, Calvin's Board of Trustees, which then consisted largely of clergymen, received numerous requests to determine if Van Till

> was being faithful to the centuries-old creeds that defined Calvinism. The Trustees promptly appointed an ad hoc investigative committee. In spite of that committee's better intentions, the investigation eventually deteriorated into a repetitious probing of my commitment to the creeds specified in the Form of Subscription. While I was on a four-month retreat in Oxford, England, the committee sent me a number of their questions in writing. I responded with a carefully-crafted letter that included a candid explanation of why I questioned the requirement that members of the Calvin College Faculty sign the Form of Subscription.

Howard Van Till's publishing record shows that he has "heaped far more scorn on fellow Christian creationists and IDers (who have relatively little cultural power) than on vociferous atheists like Richard Dawkins and Daniel Dennett."[670] In spite of his new worldview, he continues to have high respect for his "former colleagues, my friends, and family members who remain loyal to the Calvinism with which I was once comfortable. They are good people, and what they hold dear I will not dismiss lightly or disrespectfully... And most of them have been exemplary in their willingness to maintain a friendly relationship with me in spite of the changes in my belief system."[671] In short, according to Van Till's own writing, he believes that God gives to the matter that He created

> all the requisite abilities to organize itself into simple living forms and then self-transform into the complex forms we see today... in the organizational and transformational capabilities of matter and

---

670 E-mail from a Darwin Doubter who has asked to remain anonymous.
671 Van Till, Howard. n.d. *From Calvinism to Freethought: The Road Less Traveled*. CFI Michigan Center for Inquiry. Presentation for the Freethought Association of West Michigan..
https://cfimichigan.org/event/from-calvinism-to-freethought-the-road-less-traveled/
http://www.freethoughtassociation.org/images/uploads/pdf/ODoRs.pdf

organisms... God is one who acts naturally, not supernaturally; who persuades, but does not coerce or overpower (naturalistic theism).[672]

This explanation amounts to an evolution core coated with a thin veneer of a mystical spirit force that he calls God.

References

Van Till, Howard J. 1986. *The Fourth Day*. Grand Rapids, MI: Eerdmans. See also: Cobb, John B (editor). 2008. Back to Darwin: A Richer Account of Evolution. Grand Rapids, MI: Eerdmans. Ch. 21 by Van Till. This book is a compilation of presentations made at the Claremont Conference where Van Till was a participant. Free Library download at: https://www.thefreelibrary.com/Back+to+Darwin%3A+A+Richer+Account+of+Evolution-a0199464281

_____. 2003. "Are Bacterial Flagella Intelligently Designed? Reflections on the Rhetoric of the Modern ID Movement." *Science & Christian Belief*, 16(2):117-140, October.

_____. 2008. Chapter 21 pp. 351-363. in John Cobb. *Back To Darwin: A Richer Account of Evolution*. Grand Rapids: Eerdmans.

_____. n.d. *From Calvinism to Freethought: The Road Less Traveled*. CFI Michigan Center for Inquiry. Presentation for the Freethought Association of West Michigan. https://cfimichigan.org/event/from-calvinism-to-freethought-the-road-less-traveled/
http://www.freethoughtassociation.org/images/uploads/pdf/ODoRs.pdf

Wilder, T. E. 1991. "The Seventh Day: Against Humanistic Biblical Interpretation." *Contra Mundum*, 1:37-46, Fall.

---

672 Van Till, 2003.

Chapter 46

Dennis Venema, Ph.D.

From Creationist to I.D. Supporter to Darwinist

Dennis Venema is currently Associate Professor of Biology at Trinity Western University in Canada.[673] This case is important because Venema does not admit to being an atheist. If he did he would lose his college position. There is, however, no doubt in my view that he will influence many of his students to eventually accept atheism as has occurred at other Christian colleges.

Venema grew up in northern British Columbia, Canada, where he spent much time with his father and brother hunting and fishing. As a child, he remembers being very interested in how nature worked. While his peers wanted to be astronauts and firemen, he dreamed of being a scientist. Creation science was largely ignored in his local church, but rather, it seemed global missions was their priority. As such, science–faith issues were seldom, if ever, discussed in the church in which he grew up. According to a 2015 Pew Research Study, this is a major reason for the fact that less than half of all college age people in the U.S.A. now even claim to be Christians. The number is even lower in Canada.

Despite evolution being almost a complete non-issue in his local church, he acquired a generic anti-evolutionary position by default. He knew of no Christians who accepted it, and "can still recall the feeling of dread I would get even at hearing the word evolution spoken aloud." Even in high school biology, evolution seemed to be a non-issue and, as far as he can recall, he was never exposed to evolution then. In high school, he "found biology to be intensely boring, a mere regurgitation of information," adding that chemistry and physics seemed far more interesting, partly because they were taught based on their

> underlying principles: atomic theory, Newtonian mechanics and Einstein's theories of special and general relativity. What was missing was the theoretical underpinnings of biology: a way to organize the laundry list of information into a context. It would be a long time before I realized that evolution was the theoretical un-

---

673 This entire section was adapted from Dennis Venema, Ph.D *From Intelligent Design to BioLogos parts 1-5*. See the BioLogos web site: http://biologos.org/blog/series/from-id-to-biologos

derpinning that was missing from my biology experience. Given my dread of the topic, had this been pressed on me in high school I may have never pursued a career in biology.[674]

When still a high school student, he temporarily abandoned his childhood desire to become a scientist. In the small-town in northern Canada where he lived, he noticed a medical doctor who held a career that seemed closer to science than any career he was aware of. Accordingly, he set his sights on medicine, and in the fall of 1992, he studied biology at the University of British Columbia. One church incident that he recalls happened just before he started college. There were several recent high school graduates in his church, and

some were headed to Bible College, and others, such as myself, were off to "secular" universities. Our congregation had a time of prayer for all of us, but the contrast was stark: prayers of thanksgiving and blessing for those bible-school bound, but for those of us heading into the lion's den, prayers of supplication that we not lose our faith in the process. I can remember steeling myself for the upcoming battle, where professors tried to snare me with their atheistic teachings and peers likewise pressured me to give up my faith. One battle I knew was coming was the evolution one: certainly, as a biology student, this would be one of the challenges I would have to face.[675]

While in college, he was involved in Inter-Varsity Christian Fellowship and enjoyed the friendship of many other Christian students. Biology, however, was boring and his grades in chemistry and physics were higher than those in his declared major, biology. Evolution barely seemed to rate a mention in his coursework except in passing, and he does not recall that any compelling evidence for evolution was ever mentioned – professors seemed too intent on teaching the details of their own research, instead of focusing on information that supports evolution.[676]

Even his introductory survey courses focused more on a "description of biodiversity rather than a detailed understanding of how that diversity

---

674 This and all quotes are from his web post noted above.
675 Venema, n.d.
676 Theodosius Dobzhansky advocated that evolution should be the core of the coursework in biology. He also stressed that "nothing in biology makes sense except in the light of evolution." (*The American Biology Teacher*, 35(3):125-129). Also see "An Evaluation of the Myth That 'Nothing in Biology Makes Sense Except in the Light of Evolution,'" *Answers Research Journal*, 5:1-1. 2012.

arose." One 400-level evolution elective course was offered, which he skipped. At the beginning of his third year in college, his grades were too low to get into medical school, so he upgraded into a biology "honors" student. This meant doing a research thesis with a faculty member, and attending an "honors seminar." He writes that experiencing this

> first taste of research was electrifying: here at last was genuine science! Not long after, my upper-level classes seemed a lot more interesting and relevant, and also much easier. My grades improved dramatically, and medical school looked to be a live option once more – except for the fact that my childhood interest in science had blossomed again.[677]

His Work Against Evolution

His undergraduate thesis seminar class included requiring him to become familiar with the research of one of the biology professors at his university. One study that caught his attention was Dolph Schluter's research on experimental evolution. He writes that in class, he [Dennis] "trotted out every long-refuted, anti-evolutionary argument in the book" to defend creationism. The result was the class was fully engaged by his

> presentation, and there was some vigorous back-and-forth with some of the students who knew the science better than I because of their research work. ...The worst part was that Dolph ... was able to hear a good portion of my nonsense. Fortunately for me, Dolph had no interest in what would have been a very easy dressing-down. Rather, he restrained himself to a few words to the rest of the class on their lack of knowledge. Personally, I thought I had scored a victory for the faith, against the evils of evolution.[678]

His introduction to Intelligent Design

Not long after what he regards as this "now embarrassing" event, he was introduced to the ID movement via the writings of biochemist Michael Behe. He had "been vaguely aware of Phillip Johnson's 1991 book *Darwin on Trial*," but had at this point "not yet read any ID work in any depth." He was

---

677 Venema, n.d.
678 Venema, n.d.

> introduced to ID by a vocally pro-evolution professor who ran a lab down the hall from the lab I worked in. She maintained a … bulletin board outside her lab entitled "Crackpot's corner" that featured all manner of creationist materials. One day, an essay by Behe entitled "Molecular Machines" appeared on the board. This essay presented the argument from "irreducible complexity" that later appeared in more detail in Behe's 1996 book Darwin's Black Box.[679]

It is ironic that this "crackpot corner" making fun of creationists and ID would be posted in a so-called Christian College, the largest one in Canada! Venema read Behe's book as a new Ph.D. student in late 1996, and found it

> electrifying, and confirming in great detail what I already believed: of course evolution could not produce anything genuinely novel. … evolution was powerless to explain the intricate complexity that I knew (at a surface level) from my undergraduate biology and biochemistry classes. Of course the irreducible complexity of life called out in clear terms for a Designer, given that natural mechanisms were powerless to explain it. … Darwinism had failed and was propped up only because atheistic scientists were not willing to face the consequences of admitting the universe had a Creator.[680]

Having "sorted the issue out," he promptly shelved it because he was busy learning how to teach genetics as a Teaching Assistant, and starting his Ph.D. research – which was on fruit fly (*Drosophila*) genetics, spending hours glued to his stereomicroscope sorting anesthetized flies with a fine paintbrush. This was "science."

As a UBC graduate student, he was entitled to a library card at the neighboring Regent College seminary. Several renowned scholars (J.I. Packer, Bruce Waltke and Gordon Fee) taught there. He soon found that Regent had several decade's worth of tapes on exegesis and hermeneutics that could be checked out. He attended Gordon Fee's church then, so began with his Paul's letters tapes. Over the next few years, he would eventually exhaust the Regent library collection of Fee's tapes and moved on to Waltke and N.T. Wright among others.

---

679 Venema, n.d.
680 Venema, n.d.

One grad school activity was "journal clubs" – professors and their grad students critique a recent paper relevant to their discipline: "What one learns in this setting is invaluable … not the least that not all published papers are of equal quality." Here he experienced scientific "papers trashed for their poor experimental design and lack of appropriate controls, or vaunted for their elegance and powerful approach."

This period of his life would later influence his re-evaluation of evolution. He was learning science "for the first time as a graduate student, and also scholarly theology as well." He shares that he grew up in a church that had not prepared him for either,[681] writing that his growth in these areas at the time, did not affect his antievolutionary views, even as he learned about developmental genetics, adding the

> only effect I can recall was that I became less interested in antievolutionary apologetics in general: not that my views changed, but that I was less eager to whack folks over the head with them. As such, I fell into a "holding pattern" that would persist until after my Ph.D. and landing a job at Trinity Western. It would be at an evangelical institution where at last I would be forced to revisit my views on evolution.[682]

After graduating from University of British Columbia in 2003, he stayed on as a Post-doc to finish and publish his Ph.D. research. About this time, a position at Trinity Western University, the largest Christian College in Canada, came to his attention. He applied, was interviewed – and later found out that his extra-curricular study at Regent College was a critical factor in his landing the position. In the fall of 2004, he accepted a full-time position as Assistant Professor in Biology at Trinity.

When he was teaching college courses for the first time, and getting his fruit-fly research up and running, he notes that there wasn't time for reflection on evolution. It was a topic, however, that came up more than it had at UBC, because students at TWU were not shy about asking questions about evolution, or to consider its theological implications. Unfortunately, this is why so many Christians start down the road to atheism, or at least the class called "nones" at a Christian college. His Christian colleagues all had settled views on evolution: most in the biology department overtly accepted it, and

---

681 As I have repeatedly stressed, this is a major failure of the church and is why only about half of college age Americans now even claim to be Christians.
682 Venema, n.d.

only one environmental chemistry professor was (and is) an open ID supporter. When opportunity arose, such as when he was away at conferences, he invited this colleague to present ID to his introductory classes.

These ID presentations received positive reviews from the students, and he used them to "support my own, more subtle, intimations that evolution was an easily dismissed and highly speculative science." As such, the basic approach to evolution he "had accepted as an uncritical undergrad /early grad student continued to hold, and he saw no reason to change it."

In the fall of 2007, he had an opportunity to publish an invited paper as part of a collection of essays based on the general theme of "A Christian Perspective on…" various academic disciplines just before he was scheduled to leave for the 2007 National Association of Biology Teachers (NABT) meeting to give a paper on fruit flies. At NABT, the keynote speaker was theistic evolutionist Francis Collins, and all of the other featured presenters were connected with the *Kitzmiller vs. Dover Board of Education* trial

> in Pennsylvania a few years before: Ken Miller, biology professor at Brown University and expert witness for the prosecution, and the teachers from the Dover school who had refused to follow along with their school board's decision to teach about Intelligent Design. The court case that resulted from the board adopting a pro-ID policy tested the constitutionality of teaching ID in the US public school system, and it was a dismal failure for the ID side. I had heard nothing about the case before, nor anything about any key players in the ID movement since the mid-1990s.[683]

He admits that "so great was his ignorance" about the subject of ID, that he raised his hand in the question period, asking

> the Dover teachers if they had had any feedback from theistic evolutionists on their ordeal. They laughed, and pointed out a few upcoming sessions to be given by Ken Miller, whom I had not heard of before.[684] Later, I found a PBS NOVA documentary about the trial available as a DVD for purchase at the conference, so I

---

683 Venema, n.d.
684 Ken Miller is one of the most well-known evolutionists who claims to be a Christian, yet from my reading of his books (I have read them all) he is close to 99 percent evolutionist and appears to deny several key tenets of the Bible. He advocated the genetic chromosome fusion theory in spite of the fact that the putative fusion site is a promoter inside of a highly expressed gene, not a telomere to telomere fusion site.

bought a copy and watched it on my laptop that evening in my hotel room. It featured transcripts from the trial word-for-word as a re-enacted courtroom drama, and I was fascinated ... but I still was uncertain about where I stood on things personally.[685]

About this enormously biased and distorted PBS documentary, he writes:

> On the flight home, my head was spinning. If nothing else, *I realized that I knew virtually nothing about evolution or Intelligent Design; I had never seriously looked into either.*[686] If I was going to write anything even remotely credible on the topic, as I had now agreed to do, I had my work cut out for me. I also knew that Behe had just come out with a new book (*Edge of Evolution*), so I decided that I would start there upon my return. I had heard quite a bit of anti-ID rhetoric at the conference, and I remember thinking it best to look at the case for ID first, before looking at the case for evolution. It would turn out to be a fateful choice.[687]

After returning home from the conference, he revised his essay so extensively that only ten percent of the original essay remained. He began his research by reading Behe's then-new book *Edge of Evolution*. To give ID a chance to make its best case before he looked into the evidence for evolution, he checked with his pro-ID colleague who had a copy. He closed his office door, and with note pad in hand began reading.

Rejecting His ID Beliefs

While reading those opening chapters of the book he states that

> I could hardly believe what I was reading: where was the Behe of Darwin's Black Box that had so captivated me years ago? Though it is not polite to recount it ... I clearly recall putting EoE [*Edge of Evolution*] down on my desk thinking, "What is this?" I was shocked: I had fully expected to once again be amazed and amused watching Behe take evolution down a peg or two. Yet here I was, *knowing virtually nothing of evolution*, and already I was

---

685 Venema, n.d.
686 He is being honest here. Remember, he has a Ph.D. in genetics and was a biology professor at the largest Christian college in Canada.
687 Venema, n.d.

seeing nothing but holes in Behe's argument. Later on, when Behe began to discuss a topic I was familiar with (population genetics) I confirmed what I suspected: Behe was out of his area of specialty and out of his depth.[688] Later work would convince me that this pattern applied to the whole of the book and the core of Behe's arguments.[689]

Before he had even finished reading *Edge of Evolution* (how much he read he did not say), he

was done with ID. I would lose my faith in ID not by comparing it to the science of evolution, but by reading one of its leading proponents and evaluating his work on its own merits. ID, I decided, was an argument from analogy, ignorance and incredulity. I was looking for an argument from evidence. Due to an interesting set of circumstances, I was able to read Behe both as a credulous lay reader and as a skeptical trained scientist.[690] Behe, I realized, hadn't changed: I had changed, and what a difference it had made.[691]

Having rejected ID, he began to look into the "evidence" for evolution. His transition happened about the same time that he rejected ID and "required only ten or fifteen minutes," to accept Darwinism, only as long as he needed to read a single research article, specifically the 2005 *Nature* paper comparing the human and chimpanzee genomes.[692] He put the article down and thought, "well, that's that," and sat back in his chair, concluding the

contrast with ID could hardly have been starker: here was nothing but argument from evidence. As a geneticist, I was fully capable of evaluating that evidence, and it was compelling. Humans and chimps were close relatives,[693] and I was no longer an anti-evolutionist … my eyes were now open to the wonder and scope of evo-

---

688 I would like to see his notes. I have read critiques by experts far more qualified than either Venema or me, and found them all wanting. The subject of this book was based in part on Behe's Ph.D. thesis at Pennsylvania State University.
689 Venema, n.d.
690 Yet he admits that he "knew virtually nothing about evolution or Intelligent Design; I had never seriously looked into either."
691 The fact is, the likelihood of two or more mutations occurring that produce an evolutionary beneficial change is very unlikely, three are close to impossible given the fact that 99.9 percent of all mutations are near neutral or deleterious. Dennis provides no evidence to dispute this fact.
692 Tomkins and Bergman. 2012. See Waterson, et al. 2015.
693 In fact, there exists a genetic chasm between chimps and humans so great that even creationists cannot understand the reason why the great difference exists between the two genomes.

lution as a foundational theory of biology: everywhere I looked, evolution informed what I knew, whether in cell biology, genetics, immunology or developmental biology. In an instant, the pieces clicked together, and I reveled in the deeper understanding.[694]

My response

It is amazing that one can change their entire worldview just by reading one short article! Ironically, this article has been carefully refuted. In "Is the human genome nearly identical to chimpanzee?—a reassessment of the literature," he claims that "we see 99.4% identity for the entire coding sequence, [of humans and chimps] not just in these areas.[695] Even if the data that Venema cites was accurate, and the difference is only 1%, this difference translates into 30,000,000 base-pairs difference between the two genomes. Venema adds that "the commonly-agreed value of 96-99%,"[696] means from 30,000,000 to 120,000,000 differences. One major published study concluded: "The difference between the two genomes is actually not 1%, but 4%—comprising 35 million single nucleotide differences and 90 Mb" (Comparing the human and chimpanzee genomes: Searching for needles in a haystack) of insertions and deletions.[697]

The problem is, the entire genome of the chimps and humans has not yet been sequenced, nor has a viable comparison ever been made, partly because it is difficult to compare data sets that are so different. An example is, the difference is likely closer to 85 percent, or almost .5 billion base pair differences. It took me almost 20 years of study to realize that molecules to man evolution could never have occurred, and did not occur, a conclusion based on science and not theology, and Venema rejected ID in a few days.[698]

Venema then writes that, as the essay took shape, he was able to incorporate his new worldview in his paper, noting he found two books especially helpful: *Finding Darwin's God* by Ken Miller, and *The Language of God* by

---

694 Venema, n.d.
695 For a few examples see Jerry Bergman and JeffreyTompkins. 2012. *Journal of Creation*. 25(4):54–60. And JeffreyTomkins and Jerry Bergman. 2012. "Genomic monkey business—estimates of nearly identical human-chimp DNA similarity re-evaluated using omitted data," *Journal of Creation*. 26(1):94-100. April.
696   http://biologos.org/blog/adam-eve-and-human-population-genetics-addressing-criticspoythress-chimpanz
697 Varki & Altheide, 2005.
698 Carter, 2017. An excellent review of Venema's arguments against Adam as the first man are found here.

Francis Collins.[699]

> Though colleagues at TWU counseled against being "too open" about my new views, I was determined that the essay reflect what I thought to be the best way to put science and faith together. In the end, the essay would receive positive reviews despite its embrace of evolution as one of God's creative mechanisms, and its lack of support for the Intelligent Design movement, Young-Earth creationism or Old-Earth creationism.[700]

My brief response to Collins' book *The Language of God: A Scientist Presents Evidence for Belief* is as follows:

> The chapter on Intelligent Design was irresponsible. Just three of the many examples that I noted are, first, the common claim that we have back problems because our back is "poorly designed." This once common belief caused untold suffering. Patients were told to treat most common back problems with pain medicine, bed rest, or surgery.
>
> Now the common solution is walking, exercise, activity and avoid sitting for hours at a time with poor posture. The problem is not that we evolved from some primate ancestor who walked on all fours, but because our back is not designed for modern sedentary life but for walking. Verna Wright, Co-director of Bioengineering at Leads University, calls the claim that upright posture is the major culprit for back problems in humans "nonsense."
>
> Collins gives no evidence to support his claim that "the human spine [is] not optimally designed for vertical support." Having studied the spine in medical school, I am amazed the abuse that it can take for years and yet still function for decades with few problems. After looking extensively in the literature, I have yet to see any evidence that the existing design is not optimal. Conversely, many animals that walk on all fours, such as dogs, have major back problems. German Shepherds and or dachshunds are common examples.
>
> The second irresponsible and common claim that Collins refers to is the human eye, which he concluded, "is not completely

---

699 These books have been carefully refuted by many observers. For example, see A review of Finding Darwin's God by Kenneth R. Miller in *Journal of Creation*, 15(3):29–35. December 2001.
700 Venema, n.d.

ideal." In fact, there are very good reasons to conclude that the existing inverted eye design is far better than ideal. One major reason for the retina reversal is because it allows both the rods and cones to closely associate with retinal pigment epithelial cells that provide the retina with nutrients, recycle photopigments, and provide an opaque layer to absorb excess light. This design is superior to other designs because it allows the rods and cones to have the intimate association with the pigmented epithelium required to maintain the photoreceptors. Rods and cones require an enormous energy level, both for the very high metabolism rate required for them to function, and for their maintenance and repair.

In addition, because of phototoxicity damage, the rods and cones must completely replace themselves every seven days or so. The nerves in front of the rods and cones are almost totally transparent, so have little effect on light travel. Human-designed cameras still have a long way to go before they will equal the effectiveness of the human eye. The sensitivity of the existing human inverted eye design is so great that only a single photon is able to elicit an electrical response. One can hardly obtain a design superior to this.

The claim that the appendix is useless is also irresponsible. I have taught anatomy for nearly 30 years to health science majors, and not a single anatomy textbook that we have reviewed or used has claimed this. All correctly noted its important immunological and bacteria safe-house design to ensure that, after the commensal bacteria in the gut are wiped out due to diarrhea, or the use of antibiotics, the good bacteria are replaced. As an M.D., Collins should know all of these facts.[701]

Venema adds that soon additional opportunities would exist "for engaging science-faith issues. What I had previously largely avoided was now an area of interest, and a natural fit for both my training in the sciences and my commitment to evangelical Christianity." He believed that he now also had an "an opportunity to make amends for a previous mistake," namely his opposition to evolution. As he related, his "transition from aligning myself with the Intelligent Design Movement to accepting evolution was rather sudden" and the few factors which had helped included his training in biology. The

---

701 Bergman, Jerry, 2006. Review of Francis Collin's book *The Language of God*. November 1. Unpublished paper.

problem is, he argues, most "evangelicals cannot read the primary scientific literature on evolution as part of their own journey" whereas Venema claims he could. Further, helpful in his journey to an evolutionist was, ironically, the

> theological material that I had spent years listening to as a graduate student. Through that material I had learned that the simple, straightforward, Sunday-school approach to the Bible that I had learned as a child and teenager was merely a façade: Scripture was interwoven with mystery, tensions and scholarly issues that are simply not discussed in the average evangelical church. Though many pastors learn about these issues in seminary, most will never mention them from the pulpit for fear of unsettling the faith of their congregations.[702]

Venema concluded that he now has washed away his "tendencies of rigid thinking" and knew that the first "chapters of Genesis had the hallmarks of an ancient near-eastern worldview." In other words, it is a myth copied from the pagan cultures existing around the Hebrews. Venema then concluded that

> evolution, including human evolution, as a well-supported scientific theory did not precipitate a theological crisis for me. Ironically, what many pastors fear to touch in a Sunday morning sermon was just what I needed to handle this shift. This did not mean, of course, that I had everything worked out theologically then (or that I do now). Rather, it had created habits of mind that were more at ease with exploring uncomfortable questions, and reevaluating long-held assumptions.[703]

Specifically, he came to realize that, in contrast to his "relatively conservative church experience," his "relationship with God was not tied to a specific interpretation of Genesis or a literal mode of Biblical interpretation, [and] … did not suddenly evaporate the moment I understood the evidence for our evolutionary history. Instead, God's empowering presence continued to be part of my life as I explored a method of His creative activity[704] that I had previously denied."

---

[702] Venema, n.d.
[703] Venema, n.d.
[704] Rejecting Intelligent Design indicates that God is neither Intelligent nor a Designer, thus must be ignorant and does nothing!

In 2009, during the 200th anniversary of Darwin's birth, and the 150th anniversary of the publication of Darwin's opus *Origin*, he had an opportunity to make up for his mistake of supporting Intelligent Design in the past . It was also the year that he hosted an annual meeting for professional development of biology instructors from universities and colleges all over British Columbia. Accordingly, he needed to arrange a plenary speaker. The theme was the 200th anniversary of Darwin's birth, so Venama used as the speaker Dolph Schluter, internationally-known for his work on the freshwater stickleback fish. Multiple coastal lakes in British Columbia that were colonized with marine sticklebacks make Dennis' home province a natural laboratory for adaptive radiation.

Dolph's research was also once the target of his anti-evolutionary views as an undergraduate, some twelve years prior, thus he felt Dolph would be perfect in more ways than one. As he introduced him to the crowd of faculty and students that attended his lecture, he recounted the story of their previous encounter and his personal transition from creationist to evolutionist. His talk, and others given on teaching evolution and interacting with students, generated much helpful discussion. He narrates that just like evolution his path to rejecting creation was sometimes slow

> and other times rapid. Small changes, whether in my thinking or in my experiences, later combined to produce larger effects. Through it all, I have no doubt that this journey was ordained and sustained by my Creator, as He patiently led me into a deeper understanding of His creation.[705] As I mentioned in a recent NPR interview, this understanding is to be welcomed, not feared. All truth is God's truth, and the book of His works is one that He desires us to take, read and celebrate.[706]

Many of these claims have been refuted in a review titled "Dennis Venema's Vacuous Arguments against ID."[707] Note that Dennis rejected both ID and the ID Designer. Even though having read only one book, and no critical commentary on it, or its arguments, or even ID arguments in general, he felt satisfied rejecting ID.

Venema explains reading Michael Behe's second book, *The Edge of*

---

705 His creation? If evolution explains everything, as per this worldview, nothing is left for God to explain.
706 Venema, n.d.
707 Cudworth, 2011. This entire section was adapted from this article and was slightly revised. http://www.uncommondescent.com/intelligent-design/dennis-venemas-vacuous-arguments

Evolution, caused him to do a complete about-face. As a biology graduate student, he had greatly admired Behe's first book, *Darwin's Black Box*, but now, as a new junior faculty member, he has decided that Behe's arguments in *The Edge of Evolution* were all wrong, and as a result, he evidently felt he needed to reject both Behe and ID.

Issues raised by Venema's account, including the argument in Behe's first book, *Darwin's Black Box*, are independent of those in the second book. *Darwin's Black Box* centers on *theoretical* difficulties for Darwinian mechanisms raised by irreducible complexity. The second book is an *empirical* argument about what Darwinian mechanisms have *in fact* accomplished in microorganisms. Even if Behe's empirical arguments in his second book were invalid, it would not follow that the arguments in his first book were also invalid. Venema never explained why he rejected the conclusions of Behe's first book on the basis of the alleged flaws in his second book.

Other questions raised by Venema's account of the alleged flaws in Behe's second book include, if Venema, by his own confession, at that time knew "virtually nothing of evolution," what made him qualified to criticize Behe's work and see the alleged "holes" in his argument?

Even if Behe was "out of his area of specialty" when discussing population genetics, that by itself does not invalidate the argument he was making. A scientist from a different field might make some errors in commenting on another field, but what needs to be shown is that the *specific slips are such as to be fatal to the argument the scientist is making.* Typical of BioLogos columnists is to make claims like: "On page 259 Meyer misnames some chemical, therefore he is scientifically incompetent, and ID is false." What must be documented is *how* misnaming some chemical, or the putative error in population genetics jargon, invalidates the entire book's argument. Venema simply makes vague, unspecified charges about Behe's incompetence without showing why the argument is invalidated by his alleged incompetence. This is the lazy man's way of arguing, and not a scientific approach.

It is very odd that Dr. Venema, a leading player in the BioLogos movement, complained about ID people writing outside of their specialties, when on the Biologos site, many columnists — Karl Giberson, Darrel Falk, Oliver Barclay, Ard Louis, and others, frequently write columns about, or make comments concerning, theology and the history of ideas — fields in which they are completely incompetent. Perhaps Dr. Venema can take his complaint about non-specialists to the BioLogos management and get something done about the theological and historical dilettantism of the scientists featured there.

Third, it does not follow that, if Behe is wrong, all ID theorists are wrong. Did Venema take the time to read the careful argument in *No Free Lunch* by William Dembski? What about *The Design of Life* by Dembski and Wells, with its careful critique of Darwinian mechanisms? Did he read the many essays by Paul Nelson, Richard Sternberg, David Berlinski, etc., which have either argued for ID or criticized Darwinism? How can he know that ID is entirely wrong based on the arguments in only one book by *one* ID proponent?

Venema wrote a series of columns on Stephen Meyer's book, *Signature in the Cell*, in an effort to refute it. Almost all of Venema's comments in those columns concerned Darwinian evolution, *which was not the topic of Meyer's book*. The topic of Meyer's book was the origin of the first life. Venema did not provide one shred of evidence that Meyer had made errors in his research and his critique of chemical evolutionary theories of the origin of life. Nor is this surprising, as Venema knows next to nothing about origin-of-life theories; his field is fruit-fly population genetics, and he has published nothing at all in the origin-of-life field.

Thus, it is understandable why he might stay away from criticizing Meyer in the specific area in which Meyer did his Ph.D. work. But Venema was unable to grasp, even when it was pointed out to him by several respondents, that his critique of Meyer was off-topic. Again, one wonders what kind of general intellectual training a scientific education provides these days when a Biology Ph.D. cannot keep his focus on the argument on the table.

Finally, if Behe was so wrong, why did Venema not publish a review of *The Edge of Evolution* in a peer-reviewed scientific journal pointing out its many flaws? Dawkins, Carroll, Coyne, Ruse and many others did. Venema made sweeping generalities about Behe's incompetence, but when it comes time to presenting the empirical evidence, he says nothing.

Dr. Venema says he was once an ardent supporter of ID. Then why has *no one in the ID movement any memory whatsoever of his support*? What conferences did he organize to bring in pro-ID speakers? What positive reviews of ID books did he write on Amazon, or in his local newspaper, or in any other venue? On what internet debating sites did he sign his name to defend ID against its critics? Where on Panda's Thumb or Pharyngula or TalkOrigins will we find his sterling defense of ID? On what platform did he debate Eugenie Scott or P. Z. Myers?

Overall, Dr. Venema's series on why he abandoned ID is much like his series of articles on *Signature in the Cell* — an intellectual washout. It contributes nothing to a serious discussion of ID arguments. If this is the best argument that BioLogos can marshal against ID, its days are numbered. Some of the publications by Dennis Venema, several written to challenge ID, include:[708]

Venema, Dennis, with Ben-Mordehai, T., and Auld, V.J.. Transient apical polarization of Gliotactin and Coracle is required for parallel alignment of wing hairs in *Drosophila*, *Developmental Biology* 275:301-314 (2004).

_____. Enhancing undergraduate teaching and research with a *Drosophila* virginizing system *CBE-Life Sciences Education* 5:353-360 (2006)

_____. "A Christian perspective on biology" in *Christian Worldview and the Academic Disciplines: Crossing the Academy*, Downey, E.D. and Porter, S.E., Eds. Eugene, OR: McMaster Divinity College Press General Series, Wipf and Stock.

_____. "Laboratory exercises to examine recombination and aneuploidy in *Drosophila*" *American Biology Teacher* 71(6):325-332 (2009).

_____. "An evangelical geneticist's critique of *Reason to Believe's* Testable Creation Model," BioLogos Foundation (2010).

_____. "Genesis and the genome: Genomics evidence for human - ape common ancestry and ancestral hominid population sizes" *Perspectives on Science and Christian Faith* 62(3):166-178 (2010).

_____. "Seeking a signature: essay book review of *Signature in the Cell: DNA and the Evidence for Intelligent Design* by Stephen C. Meyer. Perspectives on Science and Christian Faith 62(4):276-283 (2010).

_____. "Intelligent design, abiogenesis, and learning from history: A reply to Meyer" *Perspectives on Science and Christian Faith.* 63(3):183-192 (2011).

References

Bergman, Jerry. 2012. "An Evaluation of the Myth That 'Nothing in Biology Makes Sense Except in the Light of Evolution,'" *Answers Research Journal*,

---

[708] Numerous web sites respond in detail to Venema. One example is here: http://biologos.org/blogs/dennis-venema-letters-to-the-duchess/intelligent-design-and-nylon-eating-bacteria/#sthash.N9EI47Sq.dpuf
part 2: *https://www.evolutionnews.org/2017/05/the-nylonase-story-how-unusual-is-that/*
and part 3: *https://www.evolutionnews.org/2017/05/the-nylonase-story-the-information-enigma/*

5:1-1.

_____. 1993. "Evolution and the Origins of the Biological Race Theory." *CEN Tech Journal*, Vol. 7(2):155-168.

_____ and Jeffrey Tompkins. 2012. "Is the Human Genome Nearly Identical to Chimpanzee? A Reassessment of the Literature." *Journal of Creation*. 25(4):54–60.

Carter, Robert W. 2017. A review of *Adam and the Genome: Reading scripture after genetic science* (Dennis R. Venema and Scot McKnight) Journal of Creation 31(2):41-46. August

Cudworth, Thomas. 2011. "Dennis Venema's Vacuous Arguments Against ID" in *Uncommon Descent* (blog) August 18.

Tomkins, Jeffrey and Jerry Bergman. 2012. "Genomic monkey business—estimates of nearly identical human-chimp DNA similarity re-evaluated using omitted data," *Journal of Creation*. 26(1):94-100. April.

Waterson, Robert H., Eric S. Lander & Richard K. Wilson. 2005. "Initial sequence of the chimpanzee genome and comparison with the human genome," *Nature*: 437, pp. 69–87, September 1 This article has been refuted in Bergman and Thompkins, 2012.

Varki, A. & Altheide, T. K. 2005. *"Comparing the human and chimpanzee genomes: searching for needles in a haystack."* Genome Research 15, 1746–1758.

Venema, Dennis. n.d. From Intelligent Design to BioLogos parts 1-5. See the BioLogos web site
https://biologos.org/files/modules/venema_id_to_biologos.pdf

Videos

Dennis Venema: *Why I Accept Evolution (And Why You Probably Should As Well)* (2016) https://www.youtube.com/watch?v=bRGk6dqPbFE (2:05)

Dennis Venema on Adam, Eve, and Population Genetics Scot McKnight on *How Genetic Science Made Me Rethink Genesis 1-3* (2017) BioLogos Conference. https://www.youtube.com/watch?v=92NaTpS7Prc

Should Evangelicals Accept Evolution? April 7, 2016.
https://www.youtube.com/watch?v=Zs3nJICsNcQ

Human Origins and the Species Problem (27:23)
https://www.youtube.com/watch?v=PcAhBSdopTA

Chapter 47

Geerat Vermeij, Ph.D.

From Christian to Crusading Anti-Christian Atheist

Well-known Princeton graduate and Yale Ph.D., geologist Geerat Vermeij detailed his journey from Christian faith to atheism in his autobiography *Privileged Hands*. In it he noted that "as a child he found much consolation in religion," and as a young student he "was drawn to the message of hope and salvation" of Christianity because he believed that "complete faith in God and his messenger Jesus would give meaning to a lonely life."[709] As a young boy, he developed childhood glaucoma and by the age of three became completely blind. While in college he discovered evolution, and could no longer accept what he calls the rationalizations used to harmonize the Scriptures and Darwinism; Darwinism was true, the Scriptures were false. His savior was science.[710] When he quizzed others about the topic of the validity of the Bible and Genesis, he often would hear

> the unsatisfactory answer that it was a little of both. What part is true and what part is made up, I wanted to know. The answer was important to me, for I wanted to read only "true" stories, books of fact rather than fiction. If belief in God was based in part on a fictional account, elaborated by people for their own ends, I wanted no part of it.[711]

This statement documents why apologetics must go beyond scientific evidences into philosophy of science and the laws of logic. As long as the Darwinians can pull the wool over students' eyes to believe that evolution is "science" and anti-evolution is religion, they have an unfair advantage based on this questionable premise.

Professor Geerat Vermeij eventually became a committed crusader atheist and, as was true of Dr. Jones Salk, the developer of the polio vaccine, Darwinism became his god. He writes that now, as a Darwinist, evolution is the great unifying theme of biology that permeates all of his teaching from

---

709 Vermeij, 1997, pp. 27–28.
710 See Vermeij 1998. pp. 1444-1445.
711 Vermeij, 1997, p. 29.

"the design of antibiotics and the epidemiology of viral diseases to the development of language and the structure of our social system, everything about our species and about other creatures makes sense in the light of evolution."[712] He reasons that humans, although none-the-less unique in many important ways,

> are the product of natural selection, and we cannot escape evolutionary realities. Even the freedom to engage in scientific inquiry or to deny our phylogenetic link with the rest of the animal kingdom springs ... from the modification and elaboration of traits inherited from our ancestors. I believe that no educated person can remain ignorant of evolution and its all-encompassing implications.[713]

He remarks that his students have not always shared his enthusiasm for evolution, and in

> the 1970s and early 1980s, proponents of ... scientific creationism, were waging a loud campaign to have the Biblical stories of creation sanctified as science in the public schools. Each year, several students came to me proclaiming that they should not be held responsible for any evolution-inspired science I might teach, nor should they be forced to apply it on examinations.[714]

Vermeij's rabid hostility to creationism, and those who have accepted this worldview, is obvious in the following event in which he and several of his colleagues had

> agreed to take part in a public debate at Maryland in 1975 with one of the leading zealots of that movement. I choose not to reveal his name in order to perpetuate the obscurity this man so richly deserves. In front of an audience of twelve hundred people, most of them creationist partisans, our adversary trotted out the usual arguments.... When in subsequent years I received handwritten invitations to attend presentations by scientific creationists, I declined. The presence of scientists at such gatherings legitimizes the creationist cause, for it suggests as nothing else can that creation-

---

712 Vermeij, 1997, pp. 248-249.
713 Vermeij, 1997, pp. 248–249.
714 Vermeij, 1997, p. 249.

ism is an important idea worth being considered by ... evolutionary biologists.⁷¹⁵

He added that by ignoring creationists

> we rob them of credibility. Scientific creationism and its variations must indeed be fought by scientists, but on our own terms and not under rules dictated by our enemies. One such opportunity came in 1982. A bill that would require the teaching of Biblical creation as science in the public schools was introduced in Maryland's House of Delegates.... but it isn't science, and no law will make it so. Fortunately, the bill failed.⁷¹⁶

He both ignores the definition of science, and falls for the 'warfare thesis.' He is playing party politics with the Darwin Lobby, claiming that "Religion has a place but we must harness it."⁷¹⁷ One wonders who the "we" are that is going to harness it.

References

Vermeij, Geerat. 1997. *Privileged Hands: A Scientific Life.* New York: Freeman.

_____. 1998. "Fossils and the Social Future of Science." *Science*, September 4.

---

715 Vermeij, 1997, pp. 249–251.
716 Vermeij, 1997, pp. 249–251.
717 Vermeij, 1997, p. 253.

Chapter 48

Jason Wiles, Ph.D.

Ex-Creationist, Now an Evolution Propagandist

Dr. Jason Wiles was once a creationist while living in Canada until he studied evolution at McGill and Syracuse University in New York. He earned a B.A. from Harding University, an M.S. from Mississippi State University, an M.S.T. from Portland State University, and a Ph.D. from McGill University. He is now an Associate Professor in the Department of Biology at Syracuse University in New York. He lists his main interest as research in the Life and Earth Sciences, with special attention to teaching and learning about biological evolution. He teaches a class titled BIO 310: Evolutionary Biology, Religion, and Society, where he attempts to convince students that they can accept evolution and still believe in God. While many people may end up becoming theistic evolutionists, once someone subscribes to Darwinism, it is only a short step to atheism, as many of the cases in this book document.

Now a firm Darwinist due to his indoctrination into Darwinism in college, he writes that he believes "scientists have made a serious mistake in not engaging the issue,"[718] meaning they need to be indoctrinating those they interact with into Darwinism. He agrees with Susan Fisher, an entomologist at Ohio State University in Columbus, Ohio, a life-long evolutionist, who was

> shocked to learn that more than half the students in her 700-person introductory biology class identified themselves as creationists. Last year, she received funding from the John Templeton Foundation to bring in scholars, most of them Christians who reject creationism, to speak to the students. "We need to figure out among students changing their minds, what does that?"[719]

Some of Wiles' publications include the following. Note he is very active publishing articles against his former worldview.

---

718 Couzin, 2008, p. 1036.
719 Couzin, 2008, p. 1036.

Publications by Professor Wiles

Asghar, A., Wiles, J. R. & Alters, B. 2007. Canadian preservice elementary teachers' conceptions of biological evolution and evolution education. *McGill Journal of Education* 42(2): 189-209.

Asghar, A., Wiles, J. R. & Alters, B. 2007. Discovering international perspectives on biological evolution across religions and cultures. International Journal of Diversity 6(4): 81-88.

Asghar, A., Wiles, J. R., & Alters, B. 2010. The origin and evolution of life in Pakistani High School Biology. *Journal of Biological Education* 44(2), 65-71.

BouJaoude, S., Wiles, J. R., Asghar, A., & Alters, B. 2011. Muslim Egyptian and Lebanese students' conceptions of biological evolution. *Science & Education.*

BouJaoude, S., Asghar, A., Wiles, J. R., Jaber, L., Sarieddine, D., & Alters, B. 2011. Biology professors' and teachers' positions regarding biological evolution and evolution education in a Middle Eastern society. *International Journal of Science Education* 33(7).

Carter, B. E. & Wiles, J. R. 2014. Scientific consensus and social controversy: exploring relationships between students' conceptions of the nature of science, biological evolution, and global climate change. *Evolution: Education and Outreach* 7(6).

Carter, B. E., Infanti, L. M., & Wiles, J. R. 2015. Boosting students' attitudes & knowledge about evolution sets them up for college success. *The American Biology Teacher* 77(2), 113-116.

Infanti, L. M. & Wiles, J. R. (2014). "Evo in the News": A pedagogical tool to enhance students' perceptions of the relevance of evolutionary biology. *Bioscience: Journal of College Biology Teaching* 40(2), 9-14.

Infanti, L. M. & Wiles, J. R. 2014. "Evo in the News": A pedagogical tool to enhance students' perceptions of the relevance of evolutionary biology. *Teacher* 77(2):113-116; 40(2):9-14.

Snyder, J. J., Carter, B. E., & Wiles, J. R. 2015. Implementation of the peer-led team-learning instructional model as a stopgap measure improves student achievement for students opting out of laboratory. *CBE Life Sciences Education* 14(1).

Snyder, J. J. & Wiles, J. R. 2015. Peer led team learning in introductory biology: Effects on peer leader critical thinking skills. PLoS ONE 10(1):e0115084.

Wiles, J. R. 2005. Is evolution Arkansas's "hidden curriculum"? *Reports of the National Center for Science Education* 25(1-2): 32-36.

Wiles, J. R. 2006. A Threat to geoscience education: Creationist anti-evolution activity in Canada. *Geoscience Canada* 33(3): 135-140.

Wiles, J. R. 2006. Evolution in schools: Where's Canada? *Education Canada*, 46(4): 37-41.

Wiles, J. R. & Branch, G. 2008. Teachers who won't, don't, or can't teach evolution properly: A burning issue. *The American Biology Teacher* 70(1): 6-7.

Wiles, J. R. 2011. Challenges to teaching evolution: What's ahead? *Futures* 43(6).

References

Couzin, Jannijer. 2008. "Crossing the Divide," *Science* 319(5866):1034-1036. February 22.

Video: *Jason Wiles vs Lawrence Tisdall - Creationism vs Evolution – 2006* (1:15:51) https://www.youtube.com/watch?v=Lw0QVX1dLFE

Chapter 49

H. G. Wells

From Creationist to Leading Darwinist and Eugenicist

Herbert George (H.G.) Wells was one of the most well-known and important 19th- century science-fiction and science writers. Wells authored over 100 books in his long career, including such major best-selling science fiction, a genre he and Jules Verne largely invented, classics such as *The Time Machine* (1895), *The Invisible Man* (1897), *The War of the Worlds* (1898), and *The First Men in the Moon* (1901). Wells also published general fiction, and later branched out into other areas, including both history and science. Some historians claim that he changed the mind of Europe and the world on topics such as Darwinism, and for this reason Wells is called the "great sage" of the last century.[720]

His best-selling (and still in print) *Outline of History* (1920) has sold over two million copies,[721] and the four-volume *The Science of Life* (1931), for which he collaborated with his older son, George Phillip Wells and Sir Julian Huxley, also sold very well. Both *The Outline* and *Science of Life* went into great detail to defend the Darwinist worldview using many arguments now proven wrong, or foolish.

Wells' writings detail his conversion from Christian theism to Darwinism. As a youngster and Christian, Wells had a "crude conception of Evolution," but when he attended college he became fully persuaded of its validity.[722] His mother, a devout Christian, knowing that Huxley would be one of his professors, was concerned about his faith at college. Her worst fears were eventually realized. Although Wells was both impressed and influenced by Darwin's ideas in college, he at first tried to reconcile Darwinism with his faith to conform to the "simple but powerful concept, implanted by his mother's teachings when he was a child, that 'somebody must have made it all'."[723]

He later began to conclude that "there was a flaw in this assumption."[724] The evolution books that he read included Henry Drummond's *Natural Law in the Spiritual World*. Drummond, a theistic evolutionist, wrote several best-

---

720 Achenbach, 2001, p. 112.
721 West, 1984, p. 82.
722 H.G. Wells, 1934, p. 126.
723 Mackenzie and Mackenzie, 1973, p. 42.
724 H.G. Wells, 1934, p. 126.

selling books defending Darwinism and attempting to harmonize Darwinism and Christianity by adding a thin theistic veneer to his pure unadulterated Darwinism.

One important reason why Wells became an atheist was because he had a difficult time blending evolution and Christianity. As he later stated, when he embraced evolution, he could no longer accept Genesis.[725] If Genesis were false, he reasoned he could no longer accept the rest of the Scriptures. Conversely, he thought if evolution were true, the basis of Christianity, including the Fall and the sacrificial death of Christ to redeem fallen humans, were false. As a result of his reading, Wells eventually rejected both Christianity and theism. His acceptance of the "new science" of Darwinism "had dealt telling blows at revealed religion, but offered no ... alternative to it."[726]

When Wells came across a weekly atheist magazine titled *The Free Thinker*, his "worst suspicions" about Christianity appeared to him to be confirmed, and he became a committed atheist proselytizer and used his writings to spread his Darwinian gospel. He especially enjoyed the *Free Thinker's* mockeries of both religion and theism.[727] After he totally rejected theism, Wells embraced socialism and, later, even Soviet-style communism, both of which he also eventually became disillusioned with, and rejected.

Although he came from a poor family, Wells was able to study at the Normal School of Science in South Kensington under Darwin's chief disciple, Thomas Henry Huxley. He completed a first-class honors Bachelor of Science degree in zoology and another second-class honors degree in geology. His doctoral thesis from London University was titled: "The Quality of Illusion in the Continuity of the Individual Life in the Higher Metazoa with Particular Reference to the Species *Homo sapiens*." Wells began teaching college-level courses in 1891, and married his cousin Isabel the same year. The marriage did not last, partly because of Wells' openly aggressive promiscuity.

Although his mentor, T. H. Huxley, is called Darwin's Bulldog for his lifetime of tenaciously fighting for Darwinism, Wells could properly be called one of Darwin's chief apostles.[728] Huxley, Wells, and many other "eminent men of science" had an "almost fanatical faith" that science alone was the answer to "all human misery," both real and imagined.[729] Toward this end, Wells was also active in defending his new religion of Darwinism for almost

---
725 H.G. Wells, 1934, p. 127.
726 Mackenzie and Mackenzie, 1973, p. 42.
727 Mackenzie and Mackenzie, 1973, p. 43.
728 Mackenzie and Mackenzie, 1973, p. 53.
729 Mackenzie and Mackenzie, 1973, p. 55.

his entire life—a "mission, as capable of arousing enthusiasm as any religious revival."[730]

Even his fiction books actively defended Darwinism. One reviewer named Kemp concluded that Wells' novel *The Time Machine* is a "blend of Marx and Darwin."[731] This novel presents a very pessimistic view of man including the aimless evolution theme of humans becoming wretched predators who raise beautiful idiots for food. A recent[732] superbly written, well-documented 405-page biography of H. G. Wells called him one of the most important authors of the last century, both in terms of sales and influence. Wells, along with Jules Verne and Hugo Gernsback, are for good reason called the fathers of science fiction. Wells' *An Outline of History*, a book still in print, and the controversy against it by Christians, is helpful in evaluating this still popular reference work.

Professor Sherborne's biography of Wells was written partly to set the record straight about some common claims against Wells, such as the fact that he was a life-long active supporter of eugenics. Actually, with more knowledge he reversed some of his ideas in this area. Wells did actively support the so-called 'free love' movement; a lifestyle he practiced openly. Ironically, his own life eloquently illustrated the failure of the free-love lifestyle. A good third or more of the biography was about his many affairs and the tragic effect that they had on his marriages and the lives of many other persons, including the various children he fathered.

His affairs involved some of his young students, plus established writers, such as Rebecca West, Planned Parenthood founder Margaret Sanger, and quite a number of lesser known women. It was very difficult to keep up with the seemingly endless stream of lovers that Wells managed to attract, and the sordid details of his life with them, sometimes several at a time. Wells is a good subject for this book. His childhood religious education, the lack of preparation his church gave him, and the bitter fruits of his evolutionary belief stand in stark contrast to his childhood innocence.

References

---

730 Mackenzie and Mackenzie, 1973, p. 55.
731 Kemp, 1982, p. 14.
732 Sherborne, 2010.

Achenbach, Joel. 2001. "The World According to Wells." *Smithsonian*, 32(1):111-124, April.

Kemp, Peter. 1982. *H.G. Wells and the Culminating Ape; Biological Themes and Imaginative Obsessions*. London: The Macmillan Press Ltd.

Mackenzie, Norman and Jeanne Mackenzie. 1973. *H.G. Wells: A Biography by Norman and Jeanne Mackenzie*. New York: Simon and Schuster.

Sherborne, Michael. 2010. *H. G. Wells: Another Kind of Life*. London: Peter Owen.

Wells, H. G. 1934. *Experiment in Autobiography*. Boston: Little Brown.

West, Anthony. 1984. *H.G. Wells; Aspects of a Life*. London, Hutchinson.

Chapter 50

Edward O. Wilson, Ph.D.

From Born-Again Christian to Born-Again Atheist

Edward Wilson is Emeritus Professor of Entomology at Harvard University, and the author of numerous widely acclaimed books, including the now-classic text that began the new academic field called Sociobiology. Reared as a Southern Baptist "in a religious environment that favored a literal interpretation of the Bible," Wilson's loss of faith was due to his impressions of evolution as a youth and the study of evolution at the University of Alabama, where he was led to believe

> that life is connected not by supernatural design but by kinship with species having multiplied ... to create, over hundreds of millions of years, the great panoply of biodiversity around us today. If a Divine creator put it all here several thousand years ago, he also salted Earth from pole to pole with ... massive, interlocking evidence to make scientists believe life evolved autonomously.[733]

He soon realized that creationism and Darwinism were in serious conflict, deciding "that scientific materialism explains vastly more of the tangible world, physical and biological, in precise and useful detail, than the Iron-Age theology and mysticism bequeathed us by the modern great religions."[734] To him, the Darwinian worldview "offers an epic view of the origin and meaning of humanity far greater, and I believe more noble, than conceived by all the prophets of old combined. Its discoveries suggest that, like it or not, we are alone" in the universe.[735]

Robert Wright, in his book about three "brilliant" scientists, included Edward O. Wilson. In an interview about his own loss of faith, Wright wrote that both he and Wilson were reared as Southern Baptists, and both encountered the theory of evolution as teens. Wright's experience was similar to Wilson's, and his religious faith

---

[733] Wilson, 1998, p. 19.
[734] Wilson, 1998, p. 19.
[735] Wilson, 1998, p. 19.

did not survive this encounter with science in good shape. But there is one difference between Wilson and me. He seems to have had no trouble filling the void. I, in contrast, regularly get wistful about the days when the question of purpose was settled once and for all, when I knew for certain why I was here and how I was supposed to behave. And somehow I find it hard to believe that he never does. So I ask him: Doesn't he long for the days when he believed there was a God up there watching over him? Doesn't he lose any sleep over life after death?[736]

Wright proudly answered, "None," adding that:

"I don't worry about my own immortality." Still, a funny thing happened a couple of years ago. Harvard was honoring ... Reverend King, as part of the festivities, [a minister] was preaching at the Harvard Memorial Chapel. Wilson, being a southerner, was invited to the service. There was a large turnout. The reverend preached fervently, and the congregation sang richly, and one of the hymns hit home with Wilson.... Partway through it, E. O. Wilson—scientific materialist, detached empiricist, confirmed Darwinian—started crying. As if in atonement, he has a perfectly rational explanation ... "It was the feeling that I had been a long way away from the tribe."[737]

In one interview, he called himself a waffling deist, adding that "the universe made itself after it got started, however it got started."[738]

References

Wilson, Edward. 1998. *Consilience : The Unity of Knowledge*. New York : Knopf.

Wright, Robert. 1988. *Three Scientists and Their Gods: looking for meaning in an age of information*. New York: Times Books.

---

736 Wright, 1988, pp. 191–192.
737 Wright, 1988, pp. 191–192.
738 Jill Neimark. 1998. "E.O. Wilson Is on Top of the World." *Psychology Today*. September 1, p. 5 https://www.psychologytoday.com/us/articles/199809/eo-wilson-is-top-the-world

Videos

Setting Aside Half the World for the Rest of Life (56:42)
https://www.youtube.com/watch?v=7ANire8E240

E.O. Wilson Reflects on His Career and Stresses the Importance of Biodiversity (8:00)
https://www.youtube.com/watch?v=bSd2QFMoWxM

Advice to Young Scientists (14:56)
https://www.youtube.com/watch?v=IzPcu0-ETTU

On the Shoulders of Giants (59:31)
https://www.youtube.com/watch?v=DWLoluq-HCo

Diversity of Life – February 11, 2014 (1:10:25)
https://www.youtube.com/watch?v=ZV7pB5O2Txg

Chapter 51

Davis Young, Ph.D.

Creationist to Evolutionary Creationist to ????

Dr. Davis Young, a former Professor of Physics at Calvin College, went from creationist to theistic evolutionist, then, as far as I can tell, to an agnostic after he became convinced that Darwinian evolution is true. The son of a professor of the Old Testament, he was educated in public schools in Pennsylvania. He earned a B.S. with honors in geological engineering from Princeton University, an M.S. in mineralogy and geochemistry from Pennsylvania State University, and a Ph.D. in geological sciences from Brown University.

Realizing that young-Earth creationism was spreading throughout the Christian community, he set out to explain the fallacies of creationism and Flood geology to church and college audiences in articles, books, and talks. He writes that since

> retiring from Calvin College in 2004, I have resumed writing for Christian audiences, because various forms of pseudo-science doggedly persist within large segments of the church. My most recent book, *The Bible, Rocks and Time*, co-authored with my paleontologist colleague at Calvin College, Ralph Stearley, was specifically targeted at the "geology" of young-Earth creationism. We determined to root it out once and for all.[739]

A review of his book *The Bible, Rocks and Time* in *Christianity Today* complains that it is

> essentially a negative critique. Theologically, the authors seek to show that Genesis 1 need not be understood as describing six rotational days. But if so, which competing view should we adopt? They clearly dislike the "ruin-reconstruction theory" or "gap theory" (there was a large gap of time between the first and second verses of Genesis), and display reservations about the day-age view (the six days were much longer periods). The authors favor some kind of allegorical view (e.g., the "framework hypothesis"), but are

---

[739] Newcomb, Sally. 2009. "Mary C. Rabbitt History of Geology Award, Presented to Davis A. Young" Geological Society of America. www.geosociety.org/awards/09speeches/rabbitt.htm.

steadfast that they will not make a positive case for any of these. The result is that the authors do not present their own views clearly enough for critical evaluation.[740]

One of the factors that evidently motivated Young's apostasy involved the science classes he took at college. His professors taught, among other things, uniformitarian geology. Another factor was a symposium at Wheaton College that convinced him the Bible text could not be harmonized with science, and evolutionary science must take precedence. He was once a staunch advocate of the day-age theory, but he eventually abandoned this view because of the exegetical gymnastics required to harmonize Genesis with the order of events of long-age geology. He has written and published extensively for both general and specialized audiences in a number of areas. Some of his books related to this review, all of which have been carefully evaluated and found to be deficient in numerous areas, include:

Young, Davis. *Creation and the Flood: An Alternative to Flood Geology and Theistic Evolution*. Grand Rapids: Baker, 1977. Argues for a day-age, anti-evolution position.

---------- *Christianity and the Age of the Earth*. Grand Rapids: Zondervan, 1982. The book which convinced many that the Earth and life is billions of years old.

---------- *The Biblical Flood: A Case Study of the Church's Response to Extrabiblical Evidence*. Grand Rapids: Eerdmans, 1995.

---------- and Ralph F. Stearley. *The Bible, Rocks, and Time: Geological Evidence for the Age of the Earth*. Downers Grove, IL: Intervarsity Press, 2008. This is considered to be a "total rewrite" of Young's previous book titled Christianity and the Age of the Earth.

For a response to Young's view of the Flood and the age of the Earth see:

Snelling, Andrew. *Earth's Catastrophic Past: Geology, Creation & the Flood*. Dallas, TX: Institute for Creation Research, 2009. Vol. 1 and 2, 1102 pages.

Vardiman, Larry, Andrew Snelling and Eugene Chaffin. *Radioisotopes and the Age of the Earth*. Dallas TX: Institute for Creation Research, 2000.

---

740 Ross, 2009.

Reference

Ross, Marcus. 2009. "Storming Young-Earth Creationism." *Christianity Today*, April 30.
http://www.christianitytoday.com/ct/2009/april/33.63.html

Chapter 52

Frank R. Zindler

From Aspiring Minister to Militant Atheist

Reared during the late forties and early 1950s on a farm in Michigan, Frank R. Zindler aspired to be a Lutheran Minister. He took an active part in his Benton Harbor, Michigan church activities and diligently studied his catechism. Things then were going well—he even received a scholarship to attend a Lutheran seminary in Wisconsin—until his father was killed in a freak industrial electrical accident. His mother did not want Frank to leave home but, still determined to attend seminary someday, he continued to "intensively" study his Bible.[741] To help prepare himself for seminary studies, Zindler even began to study Greek and Hebrew with the help of friends that were fluent in these languages.

This all changed at Benton Harbor high school when he took a biology class. The teacher "said something that made me conclude that we would eventually cover the evil subject of evolution, a theory condemned by the Wisconsin-Synod Lutherans. Thinking it was necessary to head off the devil before he could harm the class, I resolved to refute Darwin for Jesus' sake."[742]

It turned out that the teacher never even once brought the subject of evolution up in class. Nonetheless, the aspiring minister read the *Origin of Species* from cover to cover. As a result, by the last page, he "was convinced that Darwin was right and Moses was wrong."[743] On the heels of this change in his thinking, his "church attendance tapered off rather quickly ... and the study of science became an all-consuming passion through high school." He adds that he did not leave Christianity without having "quite a number of discussions with my Lutheran Pastor" whom, evidently, had little or no training in biology, and was unable to answer his many questions.

As a result, he no longer had much respect for his pastor, or what he taught (and, evidently, what most other pastors taught as well). At the tender age of eighteen, Zindler became a firm atheist, a position from which he has not wavered since. It is apparent from reading his many writings, that Zindler has very little knowledge about the many major lethal problems with Darwin-

---

741 Zindler, 1995. p. 359.
742 Zindler, 1995. p. 360.
743 Zindler, 1995. p. 360.

ism.

He graduated from high school in the spring of 1956, maintaining the highest academic average in what was then the school's largest graduating class.[744] His chemistry scholarship allowed him to attend Kalamazoo College with the goal of becoming a physicist or a biochemist. At Kalamazoo College, although affiliated with a mainline Christian denomination, "practically everybody there was an atheist or an agnostic. Even the Dean of Chapel appeared to be a heretic." Zindler opined that "it seems that it was inevitable that I would become an atheist before graduating from college."

He completed his B.S. in Biology from the University of Michigan and, in the fall, began teaching high school biology and chemistry, but soon moved on to a better paying position in Holland, Michigan. In order to challenge the Dutch Reformed church members who attended the school where he was a teacher, he "taught biology and earth science from a strictly evolutionary point of view." No doubt, realizing that his evolution teaching implied atheism, Zindler soon was accused of teaching atheism in his classes. Nonetheless, the Holland Board of Education granted him tenure. Later, he decided to move on to other opportunities.

Awarded a graduate fellowship by the National Science Foundation, he studied molecular biology and evolution at New Mexico Highlands University.[745] Then he finished his M.S. in geology from Indiana University, and taught biology, geology, and chemistry at Fulton-Montgomery Community College. In spite of many complaints from students over his militant atheism, the college President defended him and promoted him to Chairman of the Science Division.

Although very active as an atheist, Zindler received promotions faster than most other faculty. He was even nominated by the college President for the Chancellor's Award for Teaching Excellence.[746] He became so active in his atheism, in fact, that he joined Madeline Murray O'Hair in a lawsuit to remove what he called the "religious graffiti from American currency," i.e., the phrase "In God We Trust."

During this time, Zindler also pursued doctoral studies at State University of New York in brain physiology. He later established the Ohio Chapter of American Atheists and became involved in a "dizzying whirl of debates, radio and television appearances, and speeches before the Ohio legislature."[747]

---

744 Zindler, 1995. p. 361.
745 Zindler, 1995. p. 362.
746 Zindler, 1995, p. 362.
747 Zindler, 1995, p. 363.

As a frequent contributor to various atheist publications, he wrote a regular column in *The American Atheist* called the *Probing Mind*. His other projects include what he considers his magnum opus, Inventing Jesus, a book that purported to explain "how Jesus, a purely fictional character, came to have a biography."[748] Curiously, he never wrote similar works against Mohammed or other religious leaders.

References

Zindler, Frank, 1995. pp. 358-363 in Babinski, Edward T. *Leaving the Fold: Testimonies of Former Fundamentalists*. Amherst, NY: Prometheus Books.

Video: *Frank Zindler v. William Lane Craig: Atheism v. Christianity* (5:38:37) Moderated by Lee Strobel
https://www.youtube.com/watch?v=R04eLC30FMc

---

748 Zindler, 1995, p. 363.

Chapter 53
Rev. John Zingaro
From Biblical Literalist Presbyterian Minister to Evolutionist

Rev. Zingaro was reared a Catholic and attended Catholic Schools from kindergarten to eighth grade.[749] He claimed that not once during his Catholic education did he read the Bible in school, nor was it read in Church. He relates that the first time he read a Bible was when he was age 28.[750] Now acquainted with the Scriptures, Zingaro began losing his Catholic faith in college studying journalism. He decided to join a Protestant Church, soon becoming a missionary in Africa and an enthusiastic Bible teacher.

This all changed when he was exposed to indoctrination in evolution at, of all places, a seminary. It seemed the most important thing he learned at the Presbyterian school was *not* to trust the Biblical record! He writes that one day when reading a commentary by one of his "highly respected professors … Old Testament scholar, Dr. Donald Gowan" that he was surprised to see that Gowan "wrote matter-of-factly in a commentary on Genesis that Adam and Eve had not existed as real individuals. Rather, they were merely symbolic characters representing all of humankind."[751]

In seminary, he ended up deciding that the Bible could not have been a product of God's revelation, but rather was written by people who were "swayed by the prejudices of their day."[752] As a result of this and other instruction in the seminary, his faith was shaken to the point that only a small part of Orthodox Christianity remained. By the time his seminary experience ended, Rev. Zingaro writes that he was on his way to becoming a full-fledged Darwinist.[753] This is, in fact, the topic of most of his 406-page book, *Who are the Faithful? The Struggle for Truth Against Fundamentalism*.[754] He even began to feel that the evidence which scientists "saw with their own eyes did not match the stories in the Scriptures."[755] His descent into full-fledged Darwinism included accepting the conclusion that in his words "Natural Selection is

---

749 Zingaro, 2008, p. 8
750 Zingaro, 2008, p. 12.
751 Zingaro, 2008, p. 29.
752 Zingaro, 2008, p. 128.
753 Zingaro, 2008, p. 128.
754 Zingaro, 2008, p. 83.
755 Zingaro, 2008, p. 86.

ultimately the creator of all life."

Nonetheless, he was ordained as a Presbyterian minister in 1994, and has been the pastor of the First Presbyterian Church of Newton, New Jersey since 2007. Previous to this he had served for 13 years at the Bryn Mawr Presbyterian Church in Cottage Grove, Wisconsin, a suburb of Madison. He has also spent much of his career with students, writing that he agrees with Gould, who wrote when asked if he encounters creationism as a live issue among his Harvard undergraduate students, replied

> ...only once, in thirty years of teaching, did I experience such an incident. A very sincere and serious freshman student came to my office with a question that had clearly been troubling him deeply. He said to me, "I am a devout Christian and have never had any reason to doubt evolution, an idea that seems both exciting and well documented. But my roommate, a proselytizing evangelical, has been insisting with enormous vigor that I cannot be both a real Christian and an evolutionist."[756]

Much of his book[757] is on the Dover Intelligent Design trial that he incorrectly calls a Federal Court case and erroneously claims it was about those who take the Bible literally and those who don't.[758] Actually, none of the testimony in that case revolved around the Bible, since the issue at hand was whether students should be told that materials on an alternative to Darwinism—namely Intelligent Design—were available if desired.

He quotes extensively from the trial documents in an attempt to argue that Darwinism has been proven true, ignoring the fact that serious credible opposition exists and spending much time lamenting the fact that many people reject what he calls the proven fact of Darwinism.[759] Routinely, the plaintiffs referred to their opponents as "Biblical literalists" and "religious fundamentalists," when actually the issue revolved around Intelligent Design, a scientific theory which makes no claims about religious texts or deities.

The judge in the Dover case ruled in favor of those who reject Intelligent Design. In essence, giving free rein to the atheists, Judge John E. Jones thus violated the spirit of the First Amendment of the Constitution which requires the government to be religiously neutral. The one-time Bible believing

---

[756] Zingaro, 2008, p. 151.
[757] Zingaro, 2008, pp. 161-379.
[758] Zingaro, 2008, p. 5.
[759] Zingaro, 2008, such as on p. 151.

Zingaro now talks like other dogmatic Darwinists, saying the "fundamentalists [creationists] are deluded." [760] He totally accepts the current mainline evolutionary claims, seemingly oblivious to the problems with them, and writes with apparently little or no awareness or understanding of the other side of the debate.

In summary, Zingaro's book, though lengthy, provides very few valid scientific arguments to support his conclusions about Darwinism. He reserves very little space for weighing the pros and cons of the evolutionary arguments he discusses, relying heavily on testimony of the pro-Darwin lawyers and witnesses in the Dover case. Over a decade after the case, he still quotes it extensively and uncritically, assuming that the decision of Judge Jones in one county of Pennsylvania in 2006, has decisively disproved all forms of creationism for all time. The 'Rev.' Zingaro, although he had many close female friends, later revealed that he has decided to accept the gay lifestyle. Tragically, he ended up contracting colon cancer, a disease common among sexually-active male homosexuals.[761] His book and his life is yet another well-documented example of the results of a poor grounding in apologetics and effective counter arguments to evolution which leave many unprepared for the indoctrination into secular humanism and Darwinism that is rampant not only in public schools, but in seminaries and some Christian colleges as well.

Reference

John Zingaro. 2012. *Who are the Faithful? The Struggle for Truth Against Fundamentalism.* San Bernadino, CA: Published by author.

---

760 Zingaro, 2008, p. 5.
761 Zingaro. 2008, p. 135.

Chapter 54

Joshua Zorn, Ph.D.

From Young-Earth Creationist Missionary to Anti-Creation Missionary[762]

Joshua Zorn professed faith in Christ in 1973 at the age of thirteen. It was the first time he heard that the blood of Christ shed at the Cross could wash away his sins. He immediately understood that salvation was by grace through faith, and not by works. He began a path of life in Christ, which even led him to work as a church planter in the former Soviet Union.

A few years after his conversion, he picked up a small book at a truck stop that presented Earth history from a Christian perspective. After returning home he quickly found similar literature in his local Christian bookstore and soon became an "enthusiastic devotee of young Earth creation as promoted by the Institute for Creation Research (ICR)."

Although virtually the entire academic world disagreed with the YEC view, Zorn assumed that the Biblical creationists, not they, were correct. As a budding YEC himself, he defended his views. He used to teach that evolutionists purposely distort the evidence to support their position.

Over time, as we shall discuss, he came to feel differently. He argues that pride—even if rooted in Christian belief—is the sin that most often manifests itself in attacks against "stupid atheists, secular humanists, and evolutionists" — attacks that he says he once repeatedly made himself. Now, as an evolutionist, he has publicly repented of this attitude.

College Years as a Young-Earth Enthusiast

Zorn sailed through his undergraduate years at a liberal arts college with a major in mathematics, never encountering sufficient evidence in class to shake his belief in a young Earth or his opposition to evolution. Although he perused no biology or geology classes in college, he took the initiative to hold a public lecture titled "Darwin--Was He Wrong?" to which he invited all

---

[762] All of the quotes in this section came from Dr. Joshua Zorn's paper. *The Testimony of a Formerly Young Earth Missionary*. ©1997 by the American Scientific Affiliation. http://www.asa3.org/ASA/resources/zorn.html

his friends as well as the campus at large. He had answers to what he believed were the feeble scientific objections that his fellow students could raise, which, he thought, demonstrates how very few people's beliefs are founded on facts as opposed to indoctrination. Consequently, he felt that he had carried the day in his lecture.

Up until then he had believed what ICR taught, and had not yet been exposed to other views. Feeling that a thinking person could not reject science, he was glad to observe that YEC supporters were active scientists that overthrew "philosophical naturalism which" had given rise to biblical criticism, secular humanism, and the theory of evolution. As the son of a physics professor, Zorn had a strong love for science and, as a naive and enthusiastic young believer, his mind was then fertile ground for what he now considers "ICR propaganda."

One moment of doubt occurred as an undergraduate. When walking through the university library, he noted shelf after shelf of geology books. He then asked himself "Could all these educated people be completely wrong?" By the time he entered graduate school, he had discovered geologist Davis Young's book, *Christianity and the Age of the Earth*. He had read Young's first book, *Creation and The Flood*, a few years before, and, although it sowed seeds of doubt about the young Earth, he had not yet changed his views.

As he read books supporting the old-Earth view, he became convinced that the young-Earth arguments were untenable, and learned that other Christian graduate students had problems with YEC geological arguments as well. And so, although painful, he did not want to continue to adhere to something that he now believed was wrong. Consequently, he soon rejected the young-Earth position.[763]

The Crisis

Rejection of the young-Earth view was not just a matter of science, but affected his faith and his core values. Worse yet, he was aware that, if the Earth is old, then evolution may be true. He then went through a period of deep soul-searching, which did not help him to harmonize Scripture and science. In the end, he agreed to follow what he now believed was the scientific evidence regarding the age of the Earth, but he remained open-minded, though skeptical, of the evidence for evolution. He confessed he did not have

---

763 For a response to his claims see Andrew Snelling, *Earth's Catastrophic Past. Geology:Creation & the Flood*. Dallas, TX: Institute for Creation Research. 2009. Vol 1 and 2, 1102 pages.

all the answers and, although he had read Davis Young's interpretation, did not accept it.

Evaluation of YEC Science

After Zorn abandoned his young-Earth view, for twelve years he continued to study toward a Ph.D. in mathematics with applications to population genetics. He saw argument after argument for Creation crumble in the face of the evidence he was learning in the evolutionary books he was reading—now in the hundreds—that went far beyond the age-of-the-Earth issue. The last straw for him was when the evidence forced ICR to back down on its claim of overlapping man and dinosaur tracks in the Paluxy river bed in Texas.

Some of the "man" tracks—it turns out he claims—were poorly-preserved dinosaur tracks. Since that day, he irresponsibly concluded that he no longer had any faith in any of ICR's scientific arguments, and now only rarely bothers to read their publications. He feels that: "It is unfortunate that well-meaning Christians who share both a high regard for Scripture and evangelism, have made so many scientific errors. Although it pains me to part company with Christian brethren," he now believes that ICR is doing Christianity much more harm than good. This conclusion is based on one event that occurred decades ago.[764]

Zorn also now believes that all of the YEC arguments have been refuted by both Christian and secular authors. In addition to the books by Davis Young mentioned above, he promotes those by Daniel Wonderly that he claims have refuted the YEC arguments and give additional scientific reasons to believe in an old universe and earth. He reasons that all these authors are conservative evangelicals with advanced training in science. He also recommends reading the secular critique of YEC in Kitcher's anti-Creation book, *Abusing Science*.

As a Christian, Zorn still views Scripture as authoritative, but he now believes, in the hand of a fervent believer with a certain agenda (such as the Young-Earth Creationists) Scripture can be distorted. He now believes the churches should address these issues from a more balanced perspective, meaning they should teach the old-Earth, theistic-evolution position. He assumes many pastors try to avoid controversy and, thereby, water the seeds of a spiritual crisis in the lives of YoungEarth Creationists who move on to a university.

---

[764] He didn't mention what that event was.

What Zorn concluded the Scriptures Teach

Zorn does not expect pastors or church leaders to be impressed by scientific evidence unless good hermeneutical reasons also exist for abandoning a literal reading of Genesis. As his worldview has changed, he claims to have discovered a whole new world, both of evangelical interpretations and persuasive arguments against a literalist reading of Genesis 1-3.

For example, he claims few early Jewish interpreters and church fathers held to the six consecutive twenty-four-hour day interpretation of Genesis 1. He uncritically follows Hugh Ross' claim that Philo, Justin Martyr, Irenaeus, Hippolytus, Clement of Alexandra, Origen, Augustine, Basil, and others all held to other interpretations. He fails to note some of these persons held to an instantaneous creation, other church fathers to a literal 6-day, 24-hour creation.

The 1982 summit of *The International Council on Biblical Inerrancy* debated, among other things, the ages of the universe and Earth. After deliberating papers representing various interpretations of Genesis, this group of scholars concluded belief in six consecutive twenty-four-hour creation days is nonessential to inerrancy. Everyone present except Henry Morris signed the concluding statement, thus demonstrating the isolation of ICR's position among evangelical Christians. Zorn quotes Gleason Archer, Professor of Old Testament and Semitics at Trinity Evangelical Divinity School who, concluded "Entirely apart from any findings of modern science or challenges of contemporary scientism, the twenty-four-hour theory was never correct and should never have been believed."

Incidentally, Hugh Ross' books are endorsed by several prominent theologians and Christian leaders, including Norman Geisler, Dean of Southern Evangelical Seminary; Ralph Winter, General Director of the U.S. Center for World Mission; Don Richardson, author of *Peace Child and Eternity in their Heart*; Earl Radmacher, Chancellor of Western Seminary; Walter Kaiser Jr., President of Gordon-Conwell Theological Seminary; and Stan Oaks, Director of Christian Leadership Ministries (Faculty Ministry of Campus Crusade for Christ).

Evolution

As to the argument that belief in an old Earth opens the door to belief in evolution, Zorn concluded many Christian arguments "against evolution do not stand up to scrutiny," but does not mention any examples. He believes there is still much room to doubt the scientific scenarios for the origin of life and macroevolution, although he notes many sincere Christians called "evolutionary creationists" accept most or all of evolution theory, as opposed to the naturalistic philosophy of evolution. Some Christians accept creation through evolution, but nevertheless, believe in the special creation of Adam.

The example he uses to rationalize this view is one familiar to parents. A parent telling a child that "God made you" has affirmed that God can make (create) through natural processes, i.e., sex. Zorn takes that to mean that "creation" does not require direct miraculous intervention. Instead, processes, such as mutations and selection (i.e., "evolution") can explain the existence of all life.

Zorn believes an enormous amount of scientific evidence exists for evolution which most all scientists, Christian or otherwise, accept. He says, "There are many scientific problems with the ICR position," and this is why it is not accepted by most evangelical Christians with scientific training. In such statements, though, he totally ignores the thousands of scientists that do not accept Darwinism.[765]

Christianity Not in Opposition to Science

Zorn acknowledges "the results of the historical sciences are often tentative because we cannot go back in time to directly observe what happened then." He also admits that success in locating oil deposits, an understanding of where earthquakes will occur, and the powerful argument for the existence of a Creator based on the Big Bang all depend on the accuracy of the results of the historical sciences, such as historical geology, plate tectonics and paleontology. He asserts the results of science, properly interpreted, do not challenge the authority of Scripture, but may only cause us to re-examine our interpretation of Scripture. Zorn admits there exists

> an unwarranted anti-supernatural bias in academia and elsewhere which causes many to dismiss certain Christian doctrines without a fair consideration. Christians, in reaction, tend not to trust academics and science. This bias must be exposed (see Phillip Johnson's Reason in the Balance) and opposed. As Christians we do believe in miracles, such as the resurrection of Christ, which go beyond scientific explanation. But our belief in occasional miracles is no reason for us to oppose science as such.

---

765 www.rae.org/darwinskeptics.pdf

Zorn also argued that "the worst aspect of creationism is that it creates a nearly insurmountable barrier between the educated world and the church," questioning how many persons have rejected the Gospel due to

> the unnecessary demand that converts believe that the world is no more than 10,000 years old? And how many have unnecessarily gone through a crisis of faith similar to that which he described? How many have chosen to give up their faith altogether rather than to accept scientific nonsense? How much have we dishonored our Lord by slandering scientists and their reputation?

He advocates that pastors teach what he calls "a responsible Christian viewpoint," i.e. theistic evolution, and seminaries must add basic evolutionary science courses to their requirements. Ironically, he seems oblivious to the fact that many seminaries already do this! Furthermore, he says, Christian writers should create evolutionary

> materials for Sunday school, bedtime stories, home educators, and Christian schools that will not give our children an antiscientific bias, setting them up for a crisis of faith later in life. Publishers need to have courage to publish unpopular viewpoints, if they are consistent with Christian faith. Bookstores need to be willing to sell Christian books critical of YEC that promote other views. People who are qualified to speak need to be willing to follow the Lord's call to become publicly involved, despite the persecution which will come (from well-meaning brothers in the Lord). Finally, missionaries and evangelists need to get materials expressing other viewpoints translated to oppose the virtual monopoly YEC teaching has overseas.

Of course, much of this is already happening with tragic results, as documented in this book.

When Zorn wrote this paper, he assumed YEC literature is becoming more widely distributed in the growing churches in his corner of the former Soviet Union, asserting "We are sowing the seeds of a major crisis which will make the job of world evangelism even harder than it is already." He presents no evidence that this is happening, why he alleges it is happening, or how it is likely to happen. The fact is—as this book documents—acceptance of Darwinism is much more likely to sow the seeds of doubt and unbelief that will make evangelism far more difficult than it currently is today.

Notes and Bibliography[766]

In his bibliography, Zorn provided a list of the books that he recommended because they include opinions consistent with an old Earth. Inclusion of a work in the bibliography does not imply his endorsement of all that is written in that work. In fact, as he is overseas, he has not even read some of the works listed, but included them on the strength of recommendations or book reviews. Several advocate atheism, if only indirectly. He added that an organization of mostly old-Earth Christians in the sciences is the *American Scientific Affiliation*.

References

Archer, G. "A Response to The Trustworthiness of Scripture in Areas Relating to Natural Science." In Radmacher E. and Preus R., ed. *Hermeneutics, Inerrancy, and the Bible*. Grand Rapids, MI: Academic Books, Zondervan, 1984. Gives biblical reasons why this scholar cannot accept the twenty-four-hour day interpretation of Genesis 1.

Blocher, H. *In the Beginning: The Opening Chapters of Genesis*. Downers Grove, IL: InterVarsity Press, 1984. My preferred interpretation of Genesis 13.

Grudem, W. *Systematic Theology*. Grand Rapids: Zondervan, 1994. Grudem's open-mindedness, summarized on page 308, is a good example of how someone with young-Earth tendencies may want to address the issue.

Harman, Oren. *The Price of Altruism: George Price and the Search for the Origins of Kindness*. New York: W. W. Norton, 2010.

Kitcher, P. *Abusing Science*. Cambridge, MA: MIT Press, 1982. An evolutionary critique of Young-Earth Creationism.

Lucas, E. Genesis Today: *Genesis and the Questions of Science*. London: Scripture Union, 1989. Excellent discussion of all the major issues.

Morton, G. *Foundation, Fall, and Flood*. Dallas, TX: DMD Publishing, 1995. See www.isource.net/~grmorton/dmd.htm. A very well-researched, original, and controversial attempt to harmonize Genesis and modern science.

---

[766] Zorn, Dr. Joshua. 1997. "The Testimony of a Formerly Young Earth Missionary." American Scientific Affiliation. This entire chapter was adapted from this document http://www.asa3.org/ASA/resources/zorn.html

Numbers, Ron. *The Creationists: The Evolution of Scientific Creationism.* Berkeley: University of California Press, 1992.[767]

Rademacher, E. and Preus, R., eds. *Hermeneutics, Inerrancy, and the Bible.* Proceedings from the ICBI Summit II, 1982, in Chicago, IL. Grand Rapids, MI: Academic Books, Zondervan, 1984. This work contains a key paper by Walter L. Bradley and Roger Olsen titled "The Trustworthiness of Scripture in Areas Relating to Natural Science" as well as Gleason Archer's supportive response.

Ross, H. *The Creator and the Cosmos: How the Greatest Scientific Discoveries of the Century Reveal God*, 2nd edition. Colorado Springs: Navpress, 1995. An argument for the existence of God based on recent astronomical research.

_____*Creation and Time: A Biblical and Scientific Perspective on the Creation-Date Controversy.* Colorado Springs: Navpress, 1994. Interprets the "days" in Genesis as ages. Anti-evolution. Excellent discussion of the history and tragedy of the controversy.

Rusch, Sr, Wilbert H. Origins: *What Is At Stake?* Kansas City, MO, USA: Creation Research Society, 1991.

Schaeffer, F. *No Final Conflict: The Bible Without Error in All That it Affirms.* Downers Grove, IL: InterVarsity Press, 1975. Schaeffer does not take a position on the age of the earth and claims that from a study of the Bible one could hold either opinion.

Van Till, H. J. *The Fourth Day: What the Bible and the Heavens Are Telling Us about the Creation.* Grand Rapids, MI: Eerdmans, 1986. A controversial book which claims that there is no biblical reason to oppose creation through evolutionary processes.

Wonderly, Daniel. E. *God's Time-Records in Ancient Sediments.* New York: Interdisciplinary Biblical Research Institute. Revised edition, 2013.

_____. *Neglect of Geologic Data: Sedimentary Strata Compared with Young-Earth Creationist Writings.* New York: Interdisciplinary Biblical Research Institute, 2013.

---

[767] A meticulous but biased history of the Young-Earth Creationist movement.

Chapter 55

Libby Anne

From YEC to Darwinist

Libby Anne[768] grew up in a large evangelical family that was very involved in what she now terms the Christian Right. Attending college, especially the evolutionary indoctrination she experienced, turned her world upside down. Libby today describes herself as an atheist, a feminist, and what she calls a "progressive." She blogs about leaving religion, her experience with the Christian Patriarchy and Quiver Full movements, the detrimental effects of the "purity culture," the contradictions of conservative politics, and the importance of feminism. In her words, explaining she was raised on a "line between fundamentalist and evangelical Christianity," adding she

> was homeschooled, and nearly every subject was related to God and the Bible. History was His story and our science textbooks were all [written by] creationists. My parents were great fans of Ken Ham and Answers in Genesis and I was taught to use "creation apologetics." In other words, when you evangelize someone you start by showing them the truth of young earth creationism, and after that they will have to concede the truth of the Bible and convert to Christianity.

She claims she read everything Ken Ham ever wrote (a claim that is very questionable in view of his several books and hundreds of articles), attended an Answers in Genesis conference, and even visited the Creation Museum once. Libby "was taught that we know the Bible is true because young-Earth creationism is true. As *Answers in Genesis* so often trumpets, I learned that the foundation of the Bible was a literal Genesis." Libby continues, writing that when she was in college, her "young earth creationist views were challenged" and she

> responded by fighting back. I argued with both students and pro-

---

768 Libby Anne is a pseudonym. This entire case is from the P.Z. Myers Blog known as PHARYNGULA. http://freethoughtblogs.com/pharyngula/2012/04/29/why-i-am-an-atheist-libby-anne/ and Libby Anne's blog at https://www.patheos.com/blogs/lovejoyfeminism/2012/06/raised-quiverfull-libby-anne-introductory-questions.html None of the information quoted from her is copyrighted.

fessors, sure that I had some sort of truth they were missing. I brought out every argument I had, and went back to my creationist resources for more. As time went by, though, I found my arguments effectively refuted by arguments and information I had never been exposed to before. To my utter shock, it seemed that the evidence actually fell on the side of evolution and against young earth creationism."

When she began college, Libby realized that she "did not fit in very well" and thought this was because she had been home-schooled, but it was more than that. She wore only homemade clothing, had hair down to her back, and didn't use makeup, the very definition of a fish-out-of-water in her environment. Gradually, she says, she began to make friends with evangelical girls she had met in her dorm. Libby added the God-talk was familiar to her, but none of her new friends

believed in female submission the way I did. They were in college so that they could have careers; they didn't plan to be homemakers. They were astonished when they learned that I believed I would be under my younger brother's authority if my father died, and they found my clothing and mannerisms strange and funny. Yet they accepted me as I was, and for that I will always be grateful. Without them, my transition to college would have been a great deal more painful than it was.

The dress standards she grew up with have nothing to do with creation. It appears her parents were following a sect of Christianity or the teachings of the Bill Gothard organization. It also appears she was undergoing a rebellious phase and finding some pleasure in locating arguments that undermined her parents' strict influence she subconsciously found embarrassing and restrictive. Nonetheless, she soon realized she could not effectively witness to others because she

stuck out like a sore thumb. I therefore bought myself a new wardrobe, cut my hair, and learned to wear makeup. My new clothes were still conservative, but at least they were not floor length homemade dresses. My new look worked, and I began to have theological and political conversations with a number of non-Christians. I worked hard to show them the perfection of the Bible, the evidence of young earth creationism, the evils of abortion, and the love of God.

Libby soon found a significant number of her arguments for YEC were rebutted by arguments she

> had never heard before. I was told that there were serious problems with creationism, ethical issues with the Bible, and more effective ways to decrease abortion than banning it. I turned to my resources, my books and websites on creationism, theology, and conservative politics, and I tried again. And again. And again. But some things just didn't add up.

She claims she then did "some serious research," and "was astounded" by what she learned. Specifically, she found the scientific evidence showed young-Earth creationism/Flood geology could never have happened because

> some of the rock layers I had been taught were laid down in a global flood were actually laid down in desert conditions, and in some places there are animal burrows through numerous layers. Similarly, the pollen for any given plant is only found in the layer in which that plant exists, which would make no sense if the rock layers were laid down by a huge flood. Furthermore, while I had been taught that there were no transitional fossils, I found that this was completely untrue. I was flabbergasted.

This last claim clearly illustrates the effects of indoctrination she experienced in college.[769] It also illustrated how superficial her investigation was. For example, as she reread

> and researched the Bible in order to rebut the arguments I was hearing, things stuck out to me that I had glossed over before, such as God's command that the Israelites commit genocide on neighboring tribes or the Bible's endorsement of slavery. I also began to notice errors in the Bible, such as the statement that there were 600,000 adult male Israelites among those who left Egypt, at a time when archaeological evidence shows that there were only 50,000 people living in all of Canaan. Similarly, there was no

---

[769] For a review of the other side see Andrew Snelling's *Earth's Catastrophic Past: Geology, Creation & the Flood*. Dallas, TX: Institute for Creation Research, 2009. Vol 1 and 2, 1102 pages.

> empire-wide census in the days of Caesar Augustus. I also found contradictions between the Gospels. Did Jesus ride one donkey on Palm Sunday, or two? Was he crucified at nine o'clock, or at noon? Were Mary and Joseph from Bethlehem or Nazareth? It depends on which gospel you read.

All of these concerns have been dealt with by numerous authors, but it is unlikely she would be exposed to them in the college setting. An example is

> banning abortion did not make it any less rare, but simply led to illegal abortions harmful to women. In fact, I learned that abortion is actually most rare where it is most legal – in Western Europe. The key to decreasing abortion, I found, was not picketing abortion clinics or banning it in the legislatures. It was widespread birth control.

In fact, abortion is actually very common in many places where it is legal, such as in Russia, Japan and China. Libby believed, while she had been taught to be a critical thinker, she had never questioned the beliefs her parents taught her. She concluded

> that if young earth creationism, the infallibility of the Bible, and the importance of banning abortion, things which my parents believed in so very strongly, were wrong, then everything else they had taught me was also suspect. I also realized that I could not view the Bible as I had before, as literal truth, but must instead see it as somehow figurative, spiritual, and metaphorical.

After nearly a year of this, she conceded defeat and accepted Darwinian evolution as fact. About this experience, she writes she ended up seeing everything she

> had ever known crumble at my feet. I had been taught that the truth of the Bible rested on young earth creationism. Now that that foundation was gone, I had no idea what to do with the Bible. How could I trust it? How could I believe in it? How could I interpret it? But on the other hand, how could I give it up? My entire life centered on Christ and I found my entire value in what I meant to Jesus. Without my relationship with God, my life was nothing.

Desperate to hold onto her faith, she moved away from both fundamentalism and evangelicalism toward more hierarchical and liturgical Christian tradition, such as Catholicism, searching desperately for absolute truth to "salvage what I had left of the Bible. My fascination with these older religious traditions was accompanied by a fascination in understanding where the Bible came from, who wrote it and why." She claimed she read scholarly works on this subject, but does not reveal what she read and opines that she

> saw the Bible unfolding in new and marvelous ways before my eyes as what had before been a simplistic and two-dimensional fundamentalist/evangelical understanding of the Bible deepened. At first, my reading of the history of the Bible and of the early church fathers led me to find solace in more liberal Christianity, but this solace was short lived.

The reason was, she claimed, the more she read about the Biblical record,

> the more human the book appeared. Its errors, its contradictions, and its eccentricities suddenly appeared very, very human. Yet I felt that I was being pulled in two, for I was both losing my grip with the divine and becoming incredibly fascinated with the very human development of the very human book that is the Bible ... I felt that I was being forced to choose between holding onto the divine and the beauty of total understanding.

Around this time, she read Richard Dawkins' book The God Delusion, concluding Dawkins raised questions that she had never even thought of asking and she soon

> realized that the entire center of Christianity rested on human sacrifice, [of Christ] that the Trinity was not "a mystery" but rather simply something that made no sense, and that the very idea of a hell was barbaric. I suddenly saw the God of the Old Testament as a maniacal tyrant and I realized that mankind's greatest moral achievements – such as valuing gender and racial equality and castigating human slavery – came from man, not God.

Her words here indicate her understanding of apologetics was very shallow. Then, at one point, she called a moratorium on all religious questions because she needed a time out, not time to think so much as simply time to be. At the end of the month, I turned again to questions of religion and realized that my faith had simply slipped away. It was gone… Life had gone on, and it had not lost its meaning and purpose. I still saw beauty, I still valued love, and I still had goals and dreams. And so, I closed the door on the first two decades of my life and stepped forward into the unknown.

She ends up by claiming, by teaching children their faith rests on creationism

> parents create an Achilles heel in their children. If they grow up to find that young earth creationism is wrong, they have to completely evaluate everything they believe about the Bible, God, and Christianity. In trying to buttress their children's faith, these parents build into it a fundamental flaw. Who I am today is a product of that flaw.

On her blog, she added she attended a university because college

> had always been one of my parents' expectations for me, and I've never seen them as proud as they were at my home school graduation. With my parents' approval, I chose a secular college because I wanted to witness to others and make a difference in the world. I had been taught that I was to be a culture changer, shouldn't I start now? My parents approved of this choice because they believed I was ready.

It is obvious she was not ready, a mistake parents commonly make and find that their offspring reject much of what they were taught. Libby did not reject everything she was taught, noting she believed her main role in life

> was to be a wife and mother, but no one had appeared to seek my hand and my parents, both college educated themselves, had never shaken the idea that a college degree is important. I would graduate from college, they said, and then work until someone came to my father asking for my hand, and then marry and settle down as a homemaker, wife, and mother. My plan was to find an upstanding Christian man in college and graduate with a ring on my finger. … change of heart I knew I wanted a very large family. Until then, though, I would use my college years to witness to others and further God's kingdom.

She added her family's emphasis on the importance of marriage and family actually led her to question the goal of attending college, writing she loved her life at home and loved helping her mother, asking why

> did I need to go to college when this life, here at home, was all I was meant to lead? College would not give me any additional homemaking skills, after all. Yet the expectation of college was so strong that I rarely gave voice to these questions and doubts, and instead focused on learning to keep house while at home and looked forward to the day that I would keep a house of my own.

The result of Libby's change of heart was, after rethinking her college educated parents' beliefs, she returned home

> for a semester break more worried than I have ever been in my life. What were my parents going to think about my new beliefs on evolution, the Bible, the pro-life movement, and female equality? For a few weeks I said nothing, afraid of what would happen when I did. But the longer I listened to my parents praising me for my steadfast beliefs and condemning evolution and liberal college professors the more I realized I couldn't hide my changes in belief. And so I told them. I was used to being only praised and affirmed, so telling my parents about my changing beliefs was probably the hardest thing I have ever done in my life. And sure enough, it was like I had dropped a bomb.

The result was, as has occurred thousands of times when parents sent their children to college and saw they rejected the values taught at home, she had never seen her parents more angry or disappointed then they

> were that day. I had gone from being their golden daughter to being broken, completely broken, in their eyes. With that one revelation, they learned that all of their work had been for nothing. Since their whole reason for raising me was to create a soldier for Christ, spreading their specific views around the world, my changes in belief meant that everything they had done to bring me up was wasted.

Then her

> parents' utter horror was soon replaced with attempts to retrain me

and bring me back to the straight and narrow. ... that backfired, actually, because having learned to think for myself and having seen a bit of the world, the books by the Botkins and others made no sense.

Soon thereafter, her mother repeatedly told her how much she had hurt her father, with whom she was formerly very close, and that

> if I really wanted to follow God and know what was true I should just ask my dad my questions ... But ... I had learned that my father could be, and was, wrong ... I was learning what it meant to be under authority. I was learning what it meant for my heart and mind to tell me to go one way, and my male authority to tell me to go another. And I couldn't do it. I believed too much in myself and my abilities to turn off my brain and submit to a man I no longer felt I knew. Was his love conditional? ... I wanted to yell, "I'm a *person* and I have the right to think for myself and make my own choices!" I felt suffocated, constrained. I couldn't take it. Everything was ruined and I felt that I was being asked to choose between my family and my intellectual freedom.

Rebellion might be a more accurate word. By the time Libby returned to college after her break, she had totally rejected her parents' authority, and was unsure she would return home for visits. She also paid for the rest of her college education herself by working and attending school at the same time because, she writes, her

> intellectual freedom was too important to sacrifice ... What had been so beautiful had suddenly been destroyed ... Because I had deemed to use the brain God had given me and my father had taught me to use? ... It has now been some years since I left my parents' house and shifted for myself. I think my parents were somewhat surprised that I was able to make it on my own and that I did not come home asking for help ... I found inner sources of strength ... At the same time, my college friends, both the original evangelical ones and new ones I had met, were a wonderful source of support, and always accepted me regardless of what I did or didn't believe. I finished college on my own, and was extremely proud at graduation.

During this time, not long after finishing college, she married someone who did not share her parents' beliefs and of whom her parents did not approve. Neither her parents nor siblings attended the wedding. She walked herself down the aisle. This ending indicates that some of what she was taught by her parents had stuck. She added the question if my new husband and I could visit my parents and siblings was resolved

> when we chose to become pregnant and have a child. The presence of a grandchild has improved my relationship with my parents, though it has also created new problems as they do not always agree with the way I am raising my little one. ... Regardless of the reasons for the softening of my relationship with my parents, I am grateful that I can still be a part of my siblings' lives. However, my relationship with my parents will never be the same, and the pain of what happened will never go away.

Libby feels that her parents saw her "as an empty slate and believed that they could paint on it as they wished and choose what the outcome would be. They saw me as something to be shaped and molded rather than as an individual with my own thoughts and feelings." Of course, those thoughts and feelings were heavily influenced by what she learned at college and her reading, such as the books by Richard Dawkins, especially his anti-God book *The God Delusion*, a book which, so far, has been refuted by almost a dozen books, including by a colleague of Dawkins at Oxford, Professor John C. Lennox.[770] She adds that, in the end, she chose what she called her "intellectual freedom," which, in fact, was the indoctrination she experienced at college which she accepted over the beliefs of her family, adding if she was given an opportunity she would make that same choice again because she deserves to have her

> own thoughts and feelings, my own life. And now I do. I have a wonderful husband, a sweet child, and a beautiful life. I also take pleasure in the fact that I now have excellent relationships with several adult siblings who are okay with my differences in belief. And of course, I take joy in the wonder and beauty of life unrestrained by the bonds of Christian Patriarchy.

She acknowledged her parents had taught her many good things, including

---

770 John C. Lennox 2009. *God's Undertaker: Has Science Buried God?* London: Lion Hudson, and 2011. *Gunning for God: Why the New Atheists are Missing the Target*. London: Lion Hudson.

how to care for children. Her parents even

> found a course that taught home economics, including things like balancing a checkbook and creating a budget. I learned from my mother how to shop for a large family ... As I watched my mother running the household, I was inwardly preparing myself to do the same. I am very much an organizer and a manager, and I could not wait to practice these talents in a home of my own.

Furthermore, her parents frequently hosted gatherings with other large home school families and her

> mother became something of a mentor to other home school mothers, and my father was always eager to help others. Before entertaining, we cleaned the house from top to bottom and then cooked lots of wonderful food. I loved participating in this process, enjoying both the business of preparation and the anticipation of time spent with friends.

In addition, her father

> was also a strong believer in service to those in need, and we took meals to other families, helped out with yard work, and watched other families' children when they were in a pinch. My dad was the sort of man who never says no when he sees someone who needs help, and I watched and took notes.

Her parents even hosted a Bible study that met at their house

> and we would cook and clean in preparation each week. We children loved the event, because it meant we got to see our friends, enjoy good snacks, and spent the evening playing while the adults met together. In all of these ways, hospitality was very important to my parents.

These facts from her own pen show, in fact, in many ways she had exemplary parents. This case also illustrates the importance of apologetics and how Darwinism and the rejection of creationism was the central factor that served as the door to, if not atheism, at least the rejection of her core Christian values. It also shows, although parents feel their children are ready for the indoctrination that occurs at most colleges, few 18-year-olds are.

Chapter 56

Conclusions

From the case studies I have presented in this volume, one very sobering fact emerges: far too many Christians become ensnared and led away from their relationship with Christ by the message of evolution, and their numbers are increasing. Young people entering college frequently find themselves at one or more Turning Points which challenge their most solid core Biblical beliefs. The examples presented here are only a small sample that supports the well-documented fact that Darwinism is often the doorway to atheism, or at least to a version of evolution that is covered at most by a thin veneer of theism. The challenge is to provide these students with the resources they need to effectively rebut the evolutionary claims that are so convincingly presented to them as solidly established and reliable scientific facts.

As I was editing this book, I came across two more cases. Lewis Vaughn was transformed from, in his own words, "a young Christian fundamentalist to a disillusioned agnostic to an atheist seeker of meaning in a godless world."[771] He has had a major impact on academia and is the author or co-author of over twenty books, including the textbooks *Philosophy Here and Now* (2016, 2nd ed.), Doing Ethics (2016, 4th ed.), *The Power of Critical Thinking* (2016, 5th ed.), *Anthology of World Religions* (2016), and *Bioethics* (2016, 3rd ed.). He is also the former executive editor of Free Inquiry magazine and the cofounder and former editor of Philo, a philosophy journal. As is common, his doubts were nurtured in college and Darwinism was an important part of the equation. At this point, I'd like to mention that *college professors typically promote doubt about theism, but suppress doubt about Darwinism.*

Examples Everywhere.

Other examples are the atheist and skeptic message boards which are "filled with people who grew up as creationist fundamentalists, with some of them downright bitter at the intellectual dishonesty they discovered once they got older & wiser"[772] and, the website claims, what they learned about Darwinism is allegedly a proven fact—God did not create us, but rather we were created by an accumulation of genetic mistakes called mutations manipulated

---

771 Vaughn, 2017. Introduction.
772 E-mail from William Thiery dated April 22, 2016.

by natural selection. As documented by many scientists, this view has now been falsified by empirical science.[773] It is now accepted more for social reasons than for scientific ones.[774]

The stark need is for churches, Sunday schools, Bible Colleges and especially Christian schools to close the door causing the hemorrhaging of Christians and church members defecting from the faith and good science by teaching in-depth courses in apologetics. In Christian schools, this should be done from first grade through college. Use of the material provided by various creation and ID groups will help to achieve this end. Also, they need to learn not what to think, but how to think. They need some philosophy of science, some epistemology, some logic to identify self-refuting arguments. And the cherry on top would be to study and learn the various logical fallacies employed by members of all viewpoints.[775]

Another helpful skill is to identify instances of plagiarism in atheist arguments, where they borrow Christian morality to say God is bad for ordering the slaughter of the Canaanites, etc. They need to see that one cannot sit on God's lap to slap His face. How is an evolutionist going to call something immoral? If they must "get their own dirt," they're dead. They can't evolve ANY morality by their own premises. In fact, as I have gone to great lengths to document, Darwinian morality led to the Nazis, eugenics and genocide.[776]

Doing the above will go a long way to prevent the following experience by Patrick J. Kneeling, Professor at the Canadian Institute for Advanced Research, Botany Department, University of British Columbia, in Vancouver, Canada. He was always a mediocre student and never really knew what he wanted to do for a living, and nothing seemed to excite him. This all changed in his senior year when a creationist visited his biology class.

> On that fateful day, all the science students were herded into the school auditorium, where we listened to a long and richly illustrated lecture describing literal creationism. We were informed that in an effort to "balance" our education, we would soon hear an equally long lecture on evolution. This, like many things I heard that day, turned out to be false. The evolution lecture never materialized. Remarkably, I graduated from senior biology having

---

[773] Mittledorf, Josh and Dorian Sagan. 2016. "Longevity" in *Cracking the Aging Code: The New Science of Growing Old - And What It Means for Staying Young.*
[774] Jerry Bergman. *How Darwinism Corrodes Morality: Darwinism, immorality, abortion and the sexual revolution.* 2017. Kitchener, Ontario, Canada: Joshua Press.
[775] Thou shalt not commit logical fallacies https://yourlogicalfallacyis.com/
[776] Bergman. 2011. and see also Bergman 2012a.

learned only about creationism.

School had finally gotten my full attention. I wanted to know what we were missing, and why. For the first time in my life, I willingly (eagerly even) picked up my textbook and studiously read it. With growing interest, I realized that evolution made an awful lot of sense, and that I was being hoodwinked by my biology class.

It's hard to overestimate the appeal of rebelling against the system to a teenaged boy, and that day marked the beginning of my path to a career in evolutionary biology. ...For ... one sulky teenager in the small town of Owen Sound, Ontario, it took a creationist to make him into an evolutionary biologist.[777]

In conclusion, as Joel Belz writes:

The triumph of Darwinism over the last century hasn't ultimately had much to do with science; that is only a sideshow. Darwinism's true impact has been on everything else. For if God isn't the creator of all that exists—if He isn't the first mover—then He really doesn't matter much. Now He can be marginalized. That is the implicit assumption of the media and the educational establishment throughout our society.[778]

As the introduction to this book documents, this pithy quote has been very prophetic.

To respond to the claims made by the 53 persons in this book would take many hundreds of volumes, however some responses are listed below. The key is church involvement to give their members a solid background in apologetics and material and information on both the scientific problems of evolution and the scientific evidence for creation. How to achieve this might be to make use of the following suggestions.

1. At least once a month a sermon on this issue, or, at the minimum, 2 to 4 times a year. If the minister does not feel qualified to do this, bring in a local speaker from groups such as The Institute for Creation Research (www.ICR.org), Creation Ministries International (CMI) or other creation organizations,

---

[777] Keeling, Patrick J. "Creationists Made Me Do It." *Science,* August 21, 2009, 325(5943):945.
[778] Belz, Joel. "Mind Monopolizers: Non-Christian Dogma Lies at the Root of Big Media and Education." *World*, November 11, 2000, p. 5.

including a local chapter of a creation group.

2. The minister as part of his reading program could watch videos and read a few books a year on this issue.

3. The church could encourage small groups to form that meet once or twice a month.

4. It would be a good idea for a church to acquire quality resources that can be made available to their congregation on this topic. Many films and videos are available on this issue that are designed for small group study (especially those created by Illustra Media).

5. Attend some of the many creation groups around the country that have guest speakers or presentations on various aspects of this issue. Sign up for their newsletters, YouTube channels, and alerts.

References

Bergman, Jerry 2011. *The Dark Side of Darwin*. Green Forest, AR: New Leaf Press. Worldcat lists it in 195 libraries so far. Revised edition published in 2015.

Bergman, Jerry. 2005. "Darwinism and the Deterioration of the Genome." *CRSQ*. September. 42(2):104-114.

_____, and Doug Sharp. 2008. *Persuaded by the Evidence* (editor). Green Forest, AR: Master Books.

_____. 2012a. *Hitler and the Nazis Darwinian Worldview: How the Nazis Eugenic Crusade for a Superior Race Caused the Greatest Holocaust in World History*. Kitchener, Ontario, Canada: Joshua Press.

_____. 2012b. *Slaughter of the Dissidents: The Shocking Truth About Killing the Careers of Darwin Doubters*. Volume 1, 2nd edition. Southworth, WA: Leafcutter Press.

_____. 2014. *The Darwin Effect. Its Influence on Nazism, Eugenics, Racism, Communism, Capitalism & Sexism*. Green Forest, AR: Master Books.

_____. 2015. *The Dark Side of Darwin*. Green Forest, AR: New Leaf Press.

_____. 2017. *C. S. Lewis: Anti-Darwinist: A Careful Examination of the*

Development of His Views on Darwinism. Eugene, Oregon: Wipf & Stock Publishers.

_____. 2017. *How Darwinism Corrodes Morality: Darwinism, immorality, abortion and the sexual revolution*. 2017. Ontario, Canada: Joshua Press.

Dembski, William A. and Jay W. Richards. 2001. *Unapologetic Apologetics: Meeting the Challenges of Theological Studies*, IVP Academic,

Vaughn, Lewis. 2017. *Star Map: A Journey of Faith, Doubt, and Meaning*. Farmington, MN: Freethought Books.

A Note on the Appendices

It's important to note that the information presented in the following appendices are provided without endorsement as to the veracity of all the claims made. Rather, we bring them to you to stimulate your thinking. Some sources may be more credible than others, but we feel that all of them offer some insight that will be worthy of your consideration as you work out what you believe about the issues presented here. Please also note that these resources are not intended to be comprehensive, but merely provide some perspectives and starting points we hope will be of interest as well as beneficial to readers.

K. H. Wirth, editor and publisher

Appendix A
Turning Point Topics

The following types of evidence appear to be instrumental in persuading many Christians to abandon their Biblically-based creation beliefs in favor of what appears to them to be the more credible arguments (Turning Points) offered by many scientists. Some of these Turning Points show up in the personal stories of many of the individuals presented in this book. It is hoped that these resources will assist those who seek to consider the evidence of a designed universe more carefully. The resources presented here are not intended to be complete in any sense but are instead starting points for further investigation by interested readers.

Abiogenesis (Origin-of-Life Scenarios)

This is theory that at some point life spontaneously arose naturalistically from non-living matter. Ever since the discovery of the double helix by Crick and Watson back in the 1950's there have been several attempts to demonstrate that if the building blocks of life were present, then of course evolutionary processes could work to enable the self-organization of those non-living components into living systems. Unfortunately, this has never been demonstrated in any lab anywhere in the world and so remains wishful thinking on the part of scientists. Many organic chemists and scientists in related fields have commented on the extreme unlikelihood that life could have evolved from simple building blocks as envisioned by supporters of evolution (of course many of them proclaim loudly that evolution does not address the origin-of-life issue, however that is a frail and evasive position). It's not just creationists or ID advocates who recognize the extreme unlikelihood that the first "simple" cells could have formed much less self-replicated. Scientist James Long (not an ID proponent, by the way…) has this to say:

> …even if one were given all the molecules needed in complete stereochemical purity, and the information code, could a cell be constructed using the chemical and biochemical tools that we have today? I have written about such a hypothetic experiment, and how it would be impossible, using today's expertise, to even construct the lipid bilayer, namely the exterior packaging that holds

the cell's nanomachinery in place. Just the lipid bilayer (which itself surrounds thousands of nanosystems) is beyond our ability to synthesize. The conclusion of that thought experiment is that "life based upon amino acids, nucleotides, saccharides and lipids is an anomaly. Life should not exist anywhere in our universe. Life should not even exist on the surface of the earth." "Yet we are led to believe that 3.8 billion years ago the requisite compounds could be found in some cave, or undersea vent, and *somehow or other* they assembled themselves into the first cell." If you have knowledge of chemical or biochemical synthesis, or nanosystem assembly, I encourage you to read that short article and judge for yourself. If I am wrong, then enlighten me on my error. If I am correct, then ponder how far afield we have gone in projecting to the public our knowledge of life's origin [emphasis added].

http://inference-review.com/article/an-open-letter-to-my-colleagues

He also commented that

Those who think scientists understand the issues of prebiotic chemistry are wholly misinformed. Nobody understands them. Maybe one day we will. But that day is far from today. It would be far more helpful (and hopeful) to expose students to the massive gaps in our understanding. They may find a firmer—and possibly a radically different—scientific theory. The basis upon which we as scientists are relying is so shaky that we must openly state the situation for what it is: it is a mystery." Note that since the time of my submission of that commentary cited above, articles continue to be published on prebiotic chemistry, so I will link to my short critiques of a few of those newer articles so that the interested reader can get an ongoing synthetic chemist's assessment of the proposals: [779]

http://inference-review.com/article/two-experiments-in-abiogenesis

Video: *Abiogenesis explained and discredited with Stephen Meyer.* (3:56) https://www.youtube.com/watch?v=rWSCpwKQc10

Video: *Origin: Probability of a Single Protein Forming by Chance.* (9:27)

---

779 James M. Tour Group. *Origin of Life, Intelligent Design, Evolution, Creation and Faith* (Updated August 2017) https://www.jmtour.com/personal-topics/evolution-creation/

https://www.youtube.com/watch?v=W1_KEVaCyaA

Abrupt Appearance

Bird, Wendell R. 1989. *The Origin of Species Revisited: The Theories of Evolution and of Abrupt Appearance* (2 vols.) New York, NY: Philosophical Library.

Video: Darwin's Doubt (1:25:47) https://www.youtube.com/watch?v=QiDmtDuMHSc (start at 11:45 into the video) or get the book by the same title written by Stephen C. Meyer, where he examines the mysterious and sudden appearance of life in the Cambrian explosion.

Age of the Earth

Video: Intelligent Design and the Age of the Earth by Stephen Meyer (9:03) https://www.youtube.com/watch?v=nn4KFzX6Ywg

Video: Astronomy Reveals a 6,000 Year Old Earth by Dr. Jason Lisle https://www.youtube.com/watch?v=bEejUyEyhhg&t=14s

DeYoung, Donald. *Thousands ... Not Billions: Challenging an Icon of Evolution, Questioning the Age of the Earth*. Green Forest, AR: Master Books, 2005, 190 pp.

Humphreys, D. Russell (1979). *Starlight and Time: Solving the Puzzle of Distant Starlight in a Young Universe*. Colorado Springs, Colorado: Master Books, 1994, 133 pp. Humphreys argues for a recent creation of the Earth using a known derivation of relativity and a fresh insight into the information usually used to support Big Bang cosmology. Includes a section of the scientific basis for his theory, the biblical basis, previous theories and a reprint of several technical papers which argues for his worldview. Humphreys is a physicist at Sandia National Laboratories. He has a B.S. in physics (1963, Duke University), a Ph.D. in physics (1972, Louisiana State University), and 2 patents.

_____. *Evidence for a Young World*. Florence, KY: Answers in Genesis, 2000, 24 pp.

Vardiman, Larry. *The Age of the Earth's Atmosphere*. El Cajon, CA: Institute for Creation Research, 1990. Dr. Vardiman has a B.S. in Physics from the University of Missouri at Rolla, a B.S. in Meteorology from St. Louis University and a M.S. and Ph.D. in Atmospheric Science from Colorado State University.

_____. *Ice Cores And the Age of the Earth.* El Cajon, CA: Institute for Creation Research, 1996, 73 pp. This book is a study of ice cores, how they are done, their information and the conclusions of this research on the age of the Earth.

_____. *Sea-Floor Sediment and the Age of the Earth.* El Cajon, CA: Institute for Creation Research, 1996, 94 pp. This book analyzes sea floor sediments from a catastrophic young-Earth perspective. He concludes that the evidence supports, not an old Earth but one far younger than 4.6 billion years as commonly accepted by scientists today. Includes a helpful appendix.

_____. *Over the Edge.* Green Forest, AR: Master Books, 1999, 156 pp. Defends creationism from a young-Earth view.

_____. Andrew Snelling and Eugene Chaffin. *Radioisotopes and the Age of the Earth.* Dallas, TX: Institute for Creation Research, 2000.

Age of the Universe

Video: DEBATE: Hugh Ross v. Jason Lisle – *The Age of the Universe* (1:46:29)
https://www.youtube.com/watch?v=z045s1rLLIc

Video: *Why Does the Universe Look So Old?* Albert Mohler (1:05:45)
https://www.youtube.com/watch?v=ggJZz3WkTCI

Humphreys, D. Russell (1979-). *Starlight and Time: Solving the Puzzle of Distant Starlight in a Young Universe.* Colorado Springs, Colorado: Master Books, 1994, 133 pp. Humphreys argues for a recent creation of the Earth using a known derivation of relativity and a fresh insight into the information usually used to support Big Bang cosmology. Includes a section of the scientific basis for his theory, the biblical basis, previous theories and a reprint of several technical papers which argues for his worldview. Humphreys is a physicist at Sandia National Laboratories. He has a B.S. in physics (1963, Duke University), a Ph.D. in physics (1972, Louisiana State University), and 2 patents.

_____. *Evidence for a Young World.* Florence, KY: Answers in Genesis, 2000, 24 pp.

Appearance of Age [aka Appearance of History; Functional Maturity with Immediate Youth]

Many creationists argue that the appearance of age is inherent in any living created entity and the universe. If God created life, then all life necessarily had the appearance of at least some level of maturity. There are other examples that illustrate this point in the scriptures, such as how God caused a plant to grow up over Jonah in a very short time, and how Jesus transformed water into wine without having to grow and press the grapes to make it. If we believe these accounts to be credible then we must necessarily understand that God is not limited by what He can create and that the life and universe He has created has the appearance of age.

> Then the Lord God provided a leafy plant and made it grow up over Jonah to give shade for his head to ease his discomfort, and Jonah was very happy about the plant.[780]

> [7] Jesus said to the servants, "Fill the jars with water"; so they filled them to the brim.

> [8] Then he told them, "Now draw some out and take it to the master of the banquet."

> They did so, [9] and the master of the banquet tasted the water that had been turned into wine. He did not realize where it had come from, though the servants who had drawn the water knew. Then he called the bridegroom aside [10] and said, "Everyone brings out the choice wine first and then the cheaper wine after the guests have had too much to drink; but you have saved the best till now."

> [11] What Jesus did here in Cana of Galilee was the first of the signs through which he revealed his glory; and his disciples believed in him.[781]

Morris, John D. *Did God Create with Appearance of Age?* (ICR) September 1, 1990. https://www.icr.org/article/did-god-create-with-appearance-age/

Appearance of Age: Theological Questions about Mature Creation in a Young Universe (ASA) https://www.asa3.org/ASA/education/origins/aa.htm

---

[780] Jonah 4:6
[781] John 2:7-11

Video: *Distant Starlight In a Young Universe* with David Rives and Jason Lisle (27:32) https://www.tbn.org/programs/creation-21st-century-david-rives/watch/creation-21st-century-david-rives

Appearance of Design

Many prominent evolutionists remind us that life demonstrates *what appears to be* Design, but they argue that we need to remind ourselves that this really isn't so.

"Biology is the study of complicated things that give the appearance of having been designed for a purpose."[782]

Wallace, J. Warner. "Why the Appearance of Design in Biology Is a Problem for Atheistic Naturalism". *Cold Case Christianity*, 30 October 2015. http://coldcasechristianity.com/2015/why-the-appearance-of-design-in-biology-is-a-problem-for-atheistic-naturalism/

Video: *The Secret Life of a Cell*, (Narrated with animations)

Part 1 – Organelles (3:43)
https://www.youtube.com/watch?v=Dn3eNoxQdL0

Part 1I – Organelles (3:54)
https://www.youtube.com/watch?v=aUtPUuNWCuA

Part III – The Nucleus (4:02)
https://www.youtube.com/watch?v=zwA96STHLW8

Astronomy

Video: *What You Aren't Being Told About Astronomy (Our Created Solar System)* with host Spike Psarris. How the evidence in our solar system fails to support evolutionary and cosmological assumptions. www.creationastronomy.com
Vol. I (1:50:58)
https://www.youtube.com/watch?v=CzyQbOQ0dv0
Vol. II (1:03:05)
https://www.youtube.com/watch?v=E66409i-yn4

---

[782] Dawkins, Richard. 1996. *The Blind Watchmaker*, p. 1.

Bacteria

Video: *Bacteria are not Evolving Resistance to Antibiotics* (2:06:16)
https://www.youtube.com/watch?v=_EnLhB_D4PI

Creation

Video: Debate between Ken Ham and Bill Nye: *Is Creation a Viable Model of Origins in Today's Modern Scientific Era?* (2:45:32)
https://www.youtube.com/watch?v=z6kgvhG3AkI

Video: *The Case for a Creator* with Lee Strobel
https://www.youtube.com/watch?v=ajqH4y8G0MI (59:47)

Video: *The Creation Conversation* with Stephen Meyer, Michael Behe, and David Berlinski
Part 1: (1:01:11)
https://www.youtube.com/watch?v=ZgsEtVe_Bis&t=9s
Part 2: (1:11:13)
https://www.youtube.com/watch?v=1mF7w_zF2DU

Video: *The Case for a Creator* with Jonathan McLatchie (28:32)
https://www.youtube.com/watch?v=IocsY6R-nm0

Darwinism, the Harm it has Caused

Bergman, Jerry. *Slaughter of the Dissidents*. Southworth, WA: Leafcutter Press, 2008. (Volumes 1-3 published between 2008-2018)

_____. *The Dark Side of Darwin*. Green Forest, AR: New Leaf Press, 2011. Revised edition published in 2015.

_____. 2012. *Hitler and the Nazi Darwinian Worldview: How the Nazi Eugenic Crusade for a Superior Race Caused the Greatest Holocaust in World History*. Kitchener, Ontario, Canada: Joshua Press.

_____. 2014. *The Darwin Effect. Its Influence on Nazism, Eugenics, Racism, Communism, Capitalism & Sexism*. Green Forest, AR: Master Books.

_____. 2016. *C. S. Lewis: Anti-Darwinist: A Careful Examination of the Development of His Views on Darwinism*. Eugene, Oregon: Wipf & Stock Publishers.

_____. 2017. *Evolution's Blunders, Frauds and Forgeries*. Atlanta, GA: CMI

Publishing.

_____. 2017. *How Darwinism Corrodes Morality: Darwinism, Immorality, Abortion and the Sexual Revolution.* Kitchener, Ontario, Canada: Joshua Press.

Dating Methods (confusion over and reliability of)

Video: *How Creationism Taught Me Real Science 65 Dating Assumptions* (13:58)
https://www.youtube.com/watch?v=9Ze1jO4jYE4

Carbon Dating

Video: Carbon Dating Flaws (33:49)
https://www.youtube.com/watch?v=TVuVYnHRuig

How Carbon dating works

Video: *Carbon Dating is Wrong*
https://www.youtube.com/watch?v=QbrklxLqAsw

Video: *How Does Radiocarbon Dating Work?* - Instant Egghead #28 (2:10)
https://www.youtube.com/watch?v=phZeE7Att_s

Video: *What is Carbon-14 Dating?*
Part 1: https://www.youtube.com/watch?v=svu4YlGzGkE
Part 2: https://www.youtube.com/watch?v=WTj8lWLKWKo

Video: *Doesn't Carbon 14 Dating Disprove the Bible?* (AIG) (3:43)
https://www.youtube.com/watch?v=x5BBf4EYoNc

Radiometric Dating

Video: *Radioactive and Radiocarbon Dating*
Part 1: https://www.youtube.com/watch?v=m6l-vnBvZj0 (15:25)
Part 2: https://www.youtube.com/watch?v=qc_1-4FX368 (15:18)
Part 3: https://www.youtube.com/watch?v=JshYhKpLMS8 (15:18)
Part 4: https://www.youtube.com/watch?v=LPVJbRaQkNI (15:15)

Video: *Rocks as Clocks* (26:30) with Dr. Marcus Ross
http://origins.ctvn.org/2017/09/03/rocks-as-clocks-3/

Video: *Radioactive and Radiocarbon Dating: Turning Foe Into Friend* - Dr. Andrew Snelling (1:01:44) https://www.youtube.com/watch?v=JIZo4o77kRI

Video: *Radiometric Dating* (Dr. Jason Lisle) (8:05)
https://www.youtube.com/watch?v=pR7P_6OOJCw

Video: *Radiometric Dating is Flawed* (10:17)
https://www.youtube.com/watch?v=iGDrq8rikJc

Video: *Radiometric Dating Debunked in 3 Minutes* (3:04)
https://www.youtube.com/watch?v=fg6MfnmxPB4

Video: *Radioisotope Dating—An evolutionist's best friend?* (59:35)
https://www.youtube.com/watch?v=nVvGDu9mDuQ

Video: *Do Dating Techniques Prove the Earth is Old?* Science vs. Evolution (29:53) https://www.youtube.com/watch?v=iqQVNIYhKxI

Dinosaurs

Video: *Dragons or Dinosaur* with Dr, John Morris, Darek Isaacs, Dr. David Menton, and Chuck Missler. (1:24:27)
https://www.youtube.com/watch?v=gJYqu35jtxA

Dinosaur soft tissue research

See Mark Armitage in Appendix C

Schweitzer, Mary H., Wenxia Zheng, Timothy P. Cleland, Mark B. Goodwin, Elizabeth Boatman, Elizabeth Theil, Matthew A. Marcus and Sirine C. Fakra, "A role for iron and oxygen chemistry in preserving soft tissues, cells and molecules from deep time," *Proceedings of the Royal Society B* 281:20132741, 2013

_____, Jennifer L. Wittmeyer, John R. Horner, Jan B. Toporski, "Soft-Tissue Vessels and Cellular Preservation in Tyrannosaurus rex," *Science*, March 25, 2005.

_____. 2005. "Gender-Specific Reproductive Tissue in Ratites and Tryannosaurus Rex," *Science* 3 June 2005: Vol. 308 no. 5727 pp. 1456-1460 DOI:10.1126/science.1112158

DNA

Video: *Your body's Molecular Machines.* (6:20)
https://www.youtube.com/watch?v=X_tYrnv_o6A
Animation showing cell division (mitosis) and the process of creating new strands of DNA.

Video: *The Molecular Basis of Life: Animations of Transcription, Translation, and Replication* (20:09)
https://www.youtube.com/watch?v=fpHaxzroYxg

Design in DNA – video starring geneticist Dr. Georgia Purdom (26:32)
http://origins.ctvn.org/2018/03/19/design-in-dna/
Dr. Carter has a Ph.D. in Molecular Genetics from Ohio State University. Genetics offers astonishing evidence of a Designer, who created a marvelously complex, efficient "information system" for encoding life. The complexity of DNA is problematic for molecules-to-man evolution since the necessary changes are so implausible.

Meyer, Stephen C. *Signature in the Cell: DNA and the Evidence for Intelligent Design.* New York: NY: HarperCollins, 2009.

Video: *Signature in the Cell: Stephen Meyer Faces His Critics* (1:21:02)
Part 1: https://www.youtube.com/watch?v=eW6egHV6jAw

Many people don't realize it, but Darwin did not solve, or even attempt to solve, the question of the origin of the first life. He was trying to explain how you got new forms of life from simpler forms. In the 19th century, this was a question very few scientists addressed. The standard theory in the 20th century was proposed by a Russian scientist named Alexander Oparin who envisioned a complex series of chemical reactions that gradually increased the complexity of the chemistry involved, eventually producing life as we know it. That was the standard theory, but it started to unravel in 1953 with the discovery of the structure of DNA and its information-bearing properties, and with everything we were learning about proteins and what I call the "information processing centers" in the cell, the way the proteins were processing the information on the DNA. Oparin tried to adjust his theory to account for these new discoveries, but by the mid-60s it was pretty much a spent force. Ever since, people have been trying to come up with something to replace it, and there really has been nothing that has been satisfactory. That's one of the things the book does. It surveys the various attempts and shows that in each case, the theories have a common problem: They can't explain the origin of the information in DNA and RNA. There are other problems as well, but that's the main problem. [783]

Evolution

Definitions for Evolution:

"Evolution and diversity result from the interactions between organisms and their environments and the consequences of these interactions over long periods of time. Organisms continually adapt to their environments, and the diversity of environments that exists promotes a diversity of organisms adapted to them."
Textbook: *Opportunities in Biology*, National Academics Press, 1989 (ch.8: Evolution and Diversity)

"The diversity of life on earth is the outcome of evolution: an unsupervised, impersonal, unpredictable, and natural process of temporal descent with genetic modification that is affected by natural selection, chance, historical contingencies and changing environments."
Official Position Statement: National Association of Biology Teachers, 1995.

Video: *Evolution: A Global Deception* (49:38)
https://www.youtube.com/watch?v=4wCxkBnm3ow

Video: *The Limits of Darwinism* A Presentation by Dr. Michael Behe at the University of Toronto, November 15, 2012 (1:25:33)
https://www.youtube.com/watch?v=V_XN8s-zXx4&feature=youtu.be

Mathematics of Evolution
http://www.evolocus.com/Textbooks/Hoyle1999.pdf

The Intelligent Universe
http://wasdarwinwrong.com/pdf/korthof47.pdf

Video: *Fossil Missing Link Fraud: Lucy*. With Dr. David Menton (1:11:48)
https://www.youtube.com/watch?v=QDBPIq_Ihpo

<u>*Theistic Evolution*</u>

Video: *God and Evolution: The Problem with Theistic Evolution* (1:02:35)
https://www.youtube.com/watch?v=9WHB_kMasMs

---

783 Meyer, Stephen C. Summer 2010. "Can DNA Prove the Existence of an Intelligent Designer?"? *Biola*, An interview with Stephen C. Meyer, with comments from readers.
http://magazine.biola.edu/article/10-summer/can-dna-prove-the-existence-of-an-intelligent-desi/

Video: *Did God Need Darwin?* (47:12) with Stephen Meyer
https://www.youtube.com/watch?v=mN41M732I_I

Video: *Mike Riddle on Theistic Evolution* (28:30)
Critique of BioLogos.
https://www.youtube.com/watch?v=eW6egHV6jAw

Evolution as taught in CA Public Schools

Science in the 21st Century with David Rives (II Corinthians 10:4)
Video: Dr. Dan Biddle talks about science education on evolution in CA public schools
https://www.tbn.org/programs/creation-21st-century-david-rives/watch/creation-21st-century-david-rives-38
The 10 pillars of science taught to students supporting evolutionary concepts (6:30)
Top 4 reasons (Turning Points) why students believe in evolution (9:00)

Flagellum (see also Behe resources in Appendix C – Apologists)

Video: *Bacterial Flagellum - A Sheer Wonder Of Intelligent Design* (Animation)
https://www.youtube.com/watch?v=fFq_MGf3sbk
This video shows the intricate process for self-assembly of a flagellum motor

Video: *The Amazing Flagellum*: Michael Behe and the Revolution of Intelligent Design (3:17)
https://www.youtube.com/watch?v=MNR48hUd-Hw
"This is high tech in low life"

Flood, the Great Worldwide

Video: Noah's Flood and Catastrophic Plate Tectonics (from Pangea to Today) (23:17) https://www.youtube.com/watch?v=i8SCjn1hubc&feature=youtu.be
Stunning video and graphics proposing the sequence of events for how the earth and all land life may have been destroyed by Noah's worldwide flood. A sequence of rapid and severe tsunamis are portrayed as the likely vehicle for creating sedimentary layers which resulted in the burial of massive amounts of vegetation (creating coal) and responsible for the enormous fossil boneyards found in North America and elsewhere.

Video: *Startling Evidence That Noah's Flood Really Happened* (59:25)
Lecture by Michael Oard, M.S.

https://www.youtube.com/watch?v=FGfyyozUg-8

Video: *Faith in Nature: Noah's Flood and the Development of Geology* (1:04:34)
Radcliffe Institute, Harvard University.
Includes an examination of the details, accuracy, and differences within many Flood oral traditions from different cultures as well as the history of the idea of the Flood in modern western thought. Traces the history of how the idea of a global flood was eventually dismissed before Darwin set sail on the Beagle.
https://www.youtube.com/watch?v=YMaUzNlDnSY

Video: Noah's Ark and the Flood with Dr. Georgia Purdom, PhD (1:03:01)
Answers in Genesis
Dr. Purdom explains some of the common misconceptions about the Ark and the worldwide Flood of Noah's day, and challenges the idea that the Genesis account is a myth. https://www.youtube.com/watch?v=6Ma-LP0UDtw

Video: *What Are Some Flood Evidences?* (6 Evidences) (5:18)
Answers in Genesis https://www.youtube.com/watch?v=ZGXc8Txfza4

Video: *Global Flood: Evidence from Rocks and Fossils* with Dr. Marc Surtees (54:56) https://www.youtube.com/watch?v=PXH5lwMBNiE

Video: *Origins: The Worldwide Flood, Geological Evidences*
With Dr. Andrew Snelling (geologist)
Part 1: (26:27) https://www.youtube.com/watch?v=n6JYlk9w9z0
Part 2: (26:27) https://www.youtube.com/watch?v=9wunuFmYs6o
Part 3: (26:28) https://www.youtube.com/watch?v=gXHh_lhX4jo

Video: *Explosive Geological Evidence for Creation: Mt. St. Helens* (1:25:20)
With Bruce Malone, February 20, 2007.
https://www.youtube.com/watch?v=9OIrubRBg3Q

Video: *The Receding Floodwaters: Evidence for Noah's Flood* - Michael Oard (1:24:01)
https://www.youtube.com/watch?v=2oUgux58tgc

Video: *Worldwide Evidence for Noah's Flood* (41:37) with Michael Oard
https://www.youtube.com/watch?v=jJ1QrbvBIN4

Video: *The Sedimentology of the Flood* with Dr. Kurt Wise (1:32:53)
https://www.youtube.com/watch?v=882fmumdm9A

Video: *Was There a Worldwide Flood? Four Key Pieces of Evidence* (4:40)
https://www.youtube.com/watch?v=MEWh_ztI4z8

Video: *Was Noah's Flood Local or Global?* with John Ankerberg (2:45)

https://www.youtube.com/watch?v=NyF2Bun9k9w

Video: *The Global Flood: The Reality of Noah's Ark* with Branyon May (33:12) https://www.youtube.com/watch?v=ccHOdfzSsaw

Video: *The Evidence of Creation and Noah's Flood* with Michael Rood and Kent Hovind June 12, 2018 (55:31) https://www.youtube.com/watch?v=RHh1LTEe9vk

Fossil Record

Carroll, Robert L. *Vertebrate Paleontology and Evolution*. NY: W.H. Freeman, 1988. http://doc.rero.ch/record/200124/files/PAL_E3902.pdf

Video: *The Fossil Record* with Dr. Jerry Bergman (26:31) https://www.youtube.com/watch?v=hChM__WYevc

Video: *The Evidence of the Fossil Record*. John Ankerberg hosts Dr. Stephen Meyer (28:30) https://www.youtube.com/watch?v=R4yGdJ0DOgM

Video: Darwin's Doubt (1:25:47) https://www.youtube.com/watch?v=QiDmtDuMHSc (start at 11:45 into the video)
or get the book by the same title written by Stephen C. Meyer, where he examines the mysterious and sudden appearance of life in the Cambrian explosion.

Video: *The Fossil Record: Proof of Noah's Flood or Evolution?* With Dr. Carl Warner (16:00) Genesis Apologetics (2018) https://www.youtube.com/watch?v=qHRYnm_J4ts

Video: *The Fossil Record* by Dr. Don Patton (1:09:25) https://www.youtube.com/watch?v=gSof6z_Noqs

Video: *Evolution: Challenge of the Fossil Record* with Dr. Duane Gish https://www.youtube.com/watch?v=QDBPIq_Ihpo

Video: Fossil Discontinuities: A Refutation of Darwinism and Confirmation of Intelligent Design (53:18) https://www.youtube.com/watch?v=M7w5QGqcnNs

<u>Cambrian Explosion</u>

Darwin's Dilemma (The Mystery of the Cambrian Fossil Record)
Video: https://www.youtube.com/watch?v=xxh9o32m5c0

Video: *Darwin's Doubt* with Stephen Meyer - hosted by Socrates in the City

on September 12, 2013 (1:25:47)
https://www.youtube.com/watch?v=QiDmtDuMHSc

Video: *The Evidence of the Fossil Record.* John Ankerberg hosts Dr. Stephen Meyer, 2015. (28:30) Includes commentary from others such as Paul Nelson and Jonathan Wells.
https://www.youtube.com/watch?v=R4yGdJ0DOgM

Living Fossils

Video: *Living Fossils* with Dr Carl Werner
part 1 – Origins (28:32) Dr. Werner reports on finding modern animals alongside dinosaurs in the fossil record, which is contrary to the predictions of evolution.
https://www.youtube.com/watch?v=Wz8WbuNMshI

part 2 – Origins (28:37) In this video, Dr. Werner reports on his observation about how scientists give different names to fossil representatives of living critters, rather than acknowledging that they should both have the same name. Using this "name game" approach, he suggests that this tactic is used by evolutionists to create the appearance of evolution (begin at 12:00).
https://www.youtube.com/watch?v=FZ1vjDLxSB8

For a DVD of this series, write to:
ORIGINS Program #1297
Cornerstone Television
Wall, PA 15148-1499

Send a $12 donation to cover shipping and handling.
For a PDF version of the power point presentation used in these broadcasts, you can download them for free at www.originstv.com

Video: *Living Fossils: Fossils that Debunk Evolution* (28:29)
https://www.youtube.com/watch?v=lsbRKq0tay8

Video: *What are "living fossils" and how are they a problem for Darwinian evolution?* (3:14) with Gunter Bechly
https://www.youtube.com/watch?v=5moZ5NFp-jc

Transitional Fossils

Bergman, Jerry. *Fossil Forensics: Separating Fact from Fantasy in Paleontology.* Tulsa, Oklahoma: Bartlett Publishing, 2017.

Luskin, Casey. Abrupt Appearance of Species in the Fossil Record Does Not

Support Darwinian Evolution. *Evolution News*, January 29, 2015. https://evolutionnews.org/2015/01/problem_5_abrup/

Video: *What Is the Evidence for Evolution Found in the Fossil Record?* 1:24:35
https://www.youtube.com/watch?v=S_LY-LZtJAs&t=46s
with Richard Dawkins (2009)

Video: *Richard Dawkins: Show Me The Intermediate Fossils* (2:30)
https://www.youtube.com/watch?v=DP53WAkQ5Yc
Dawkins reviews the evolution of whales.

Video: *Evolution: Where are the Transitional Forms?* (9:41)
https://www.youtube.com/watch?v=kfTbrHg8KGQ

Video: *Transitional Fossils & Missing Links: Evidence for Evolution?* (28:55)
https://www.youtube.com/watch?v=2YoAR3jkUbs

Video: *Defining Transitional Fossils* (2:12)
https://www.youtube.com/watch?v=_c6cxShUMks

Video: *The Fossil Record & Transitional Fossils: Creation or Evolution?* (47:43)
https://www.youtube.com/watch?v=jxInh3ClTTI

Video: *Transitional Fossils, Where?* (28:23)
https://www.youtube.com/watch?v=J_Oavkmvr7Q

Giants on the Earth

Video: *The Archon Invasion (Part 1)* with Ron Skiba (2:01:00)
https://www.youtube.com/watch?v=E8dUQOaSmSE

Video: *The Archon Invasion (Part 2)* with Ron Skiba (21:14)
https://www.youtube.com/watch?v=pby2Vh6AM48&t=539s

Video: *The Genesis 6 Giants* (2017) by Steve Quayle (1:48:41)
https://www.youtube.com/watch?v=oveNdIneOAE

Video: *Genesis 6 Giants* with Chuck Missler (1:12:42)
https://www.youtube.com/watch?v=0d3l2FEVmsk

Video: *Smithsonian Cover-up – Thousands of Reports of Giant Skeletons Found*
https://www.youtube.com/watch?v=HygfspmjrbM&t=2215s

Intelligent Design [784] [785]

Video: *The Evidence of Design from Biology* with Dr. Michael Behe at the University of Toronto November 16, 2012 (53:29)
https://www.youtube.com/watch?v=s6XAXjiyRfM&feature=youtu.be

Video: *Intelligent Design 3.0* – Stephen Meyer (1:36:48)
https://www.youtube.com/watch?v=lgs6J4LqeqI

Video: *The Genius of Birds* – with Dr. Paul Nelson (26:32)
http://origins.ctvn.org/2018/02/19/the-genius-of-birds-2/
Dr. Nelson has a PhD. in Philosophy of Biology from the University of Chicago.
In this video, he discusses the genius of bird design. One of the key questions he addresses is: "How can you deny the scholarly consensus that birds evolved from Theropod dinosaurs by an undirected evolutionary process?"

Video: *Intelligent Design: Is it Science?* (Debate) Lee Silver vs. William Dembski (1:55:41)
https://www.youtube.com/watch?v=iDJF4qbnbHE

Irreducible Complexity

Meyer, Stephen C., "DNA and the Origin of Life: Information, Specification, and Explanation" [date unknown] 44 pp.
www.vedicilluminations.com/downloads/Intelligent-Design/DNAPerspec-

---

[784] Michael Behe defines Intelligent Design as the "purposeful arrangement of parts."
Video: https://www.youtube.com/watch?v=s6XAXjiyRfM (at 5:46)

[785] Intelligent Design is the hypothesis that in order to explain life it is necessary to suppose the action of an unevolved intelligence. One simply cannot explain organisms, those living and those long gone, by reference to normal natural causes or material mechanisms, be these straightforwardly evolutionary or a consequence of evolution, such as an evolved extraterrestrial intelligence. Although most supporters of Intelligent Design are theists of some sort (many of them Christian), it is not necessarily the case that a commitment to Intelligent Design implies a commitment to a personal God or indeed to any God that would be acceptable to the world's major religions. The claim is simply that there must be something more than ordinary natural causes or material mechanisms, and moreover, that something must be intelligent and capable of bringing about organisms.
Intelligent Design does not speculate about the nature of such a designing intelligence. Some supporters of Intelligent Design think that this intelligence works in tandem with a limited form of evolution, perhaps even Darwinian evolution (for instance, natural selection might work on variations that are not truly random). Other supporters deny evolution any role except perhaps a limited amount of success at lower taxonomic levels – new species of birds on the Galapagos, for instance. But these disagreements are minor compared to the shared belief that we must accept that nature, operating by material mechanisms and governed by unbroken natural laws, is not enough.
Source: Dembski, William and Michael Ruse. *Debating Design*, University of Cambridge Press (2004)
https://www.cambridge.org/core/books/debating-design/general-introduction/162A7D7221A7BF6CEC18BC6B0FB147F7

tives.pdf

From the abstract: "the term 'information' can designate several theoretically distinct concepts. By distinguishing between specified and unspecified information, this essay seeks to eliminate definitional ambiguity associated with the term 'information' as used in biology."

Video: *Irreducible Complexity* with Michael J. Behe. Lecture presented at Princeton University, 2014 (1:42:47)
https://www.youtube.com/watch?v=yudmK5jZy9A

Video: *Behe & Meyer Destroy Challenge to Flagellum Motor*, 2015 (24:35)
https://www.youtube.com/watch?v=F4uJ6y5Y29g

Video: *Irreducible Complexity with Dr. Jerry Bergman*, 2017 (26:32)
https://www.youtube.com/watch?v=DxEX-itFyvY

Video: *Argument From Irreducible Complexity – Debunked*, 2017 (7:38)
https://www.youtube.com/watch?v=FS0hlXxHx78

Video: *Evolution, Design, Irreducible Complexity with Kent Hovind*, 2008 (9:09)
https://www.youtube.com/watch?v=zbO2VkWNXos

Molecular Machines

Video: *Molecular Machines and the Death of Darwinism* - Dembski, Wells, Nelson, Macosko (43:14)
https://www.youtube.com/watch?v=-D2ng2mvyes&list=PL0d1ehKrCmE3Gc2vPY58ZVfeutTBxhB3M

Video: Michael Behe - Lee Strobel - *Molecular Machines Disprove Evolution* (8:40)
https://www.youtube.com/watch?v=Y7WwO1iETuw

Video: *Molecular Motor Proteins*. (2016) Presentation by Ron Vale (2016) with animations from "Inner Life of the Cell," BioVisions: Harvard University.

Mutations (how they fail as a vehicle for evolution)

Bergman, Jerry. *The Three Pillars of Darwinism Demolished*. Dallas, TX: Institute for Creation Research, 2019.

Video: *Mutations Debunk Darwin's Evolution* with Dr. Jerry Bergman at the Seattle Creation Conference (2013). (35:44) https://www.youtube.com/watch?v=b_Acfw1RN0c

Video: David Rives & Dr. Georgia Purdom: Mutations – *A Major Problem For Evolution* circa 2018. (27:31.)

Video: *Mutations: Evolution's Disappointment* with Dr. Georgia Purdom (26:32)
http://origins.ctvn.org/2018/11/12/design-in-marine-life-3-2/

Video: *MUTATIONS – the severe problems for "Darwinian Evolution" theory.* (10:35)
https://www.youtube.com/watch?v=ovRq_xbCOXo

Video: *Why is the abrupt appearance of animal body plans a problem for the Darwinian model of evolution?* (14:38) Dr. Stephen Meyer on the John Ankerberg Show
In this clip Meyer talks about a fatal issue associated with the introduction of mutations into an organism. https://www.youtube.com/watch?v=giSGFlexOQI

Natural Selection as the Key Agent of Evolution

Brady, Ronald H. 1979. "Natural Selection and the Criteria by Which a Theory is Judged." *Systematic Zoology*, Vol. 28, No. 4 (Dec., 1979), pp. 600-621 Published by: Taylor & Francis, Ltd. for the Society of Systematic Biologists
DOI: 10.2307/2412570
https://www.jstor.org/stable/2412570
This significant article examines the tautological nature of natural selection.

Big Problems With Natural Selection – video starring geneticist Dr. Robert Carter (26:32)
http://origins.ctvn.org/2017/05/29/big-problems-with-natural-selection-2/
Dr. Carter has a Ph.D. in Marine Biology from the University of Miami
Natural Selection is often considered the very engine of evolution. Is it possible for this process to do what evolution requires?

Video: *Natural Selection is not Evolution* with Dr. Georgia Purdom (26:31)
http://origins.ctvn.org/2018/10/22/the-real-jurassic-world-2-2-2-2/

Origin of Life

Video: 2019. *The Mystery of the Origin of Life* with Dr. James Tour. (58:01)
https://www.youtube.com/watch?v=zU7Lww-sBPg&feature=youtu.be
Taped at the 2019 Dallas Science and Faith Conference at Park Cities Baptist

Church in Dallas. Sponsored by Discovery Institute's Center for Science and Culture. In this video, Dr. Tour discusses the impressive use of his work in various industries and applications. Starting at 8:00 into the video he launches into a discussion of "what does science tell us about the origin of life?" without "making any reference to God or intelligent design." He describes the complexity of cells and makes the point that molecules don't care about, assemble or evolve non-regular patterns of organic material into living systems. This video presents an excellent review of the difficulties surrounding the non-directed (i.e., evolutionary) formation of cellular components.

Origin of the Universe

Video: *Origin of the Cosmos* with Dr. Michael Strauss
https://www.youtube.com/watch?v=VhUhMiDALC0

Theistic Evolution

Video: *Science and Theistic Evolution* (with Jonathan Wells) (25:53)
https://www.youtube.com/watch?v=hksGZcqJ5h4

Video: *God & Evolution: A Critique of Theistic Evolution* (with Stephen Meyer) (2018) (2:23:43)
https://www.youtube.com/watch?v=bMMZ48M1TOo

Video: *What is Theistic Evolution?* With Stephen Meyer (4:47)
https://www.youtube.com/watch?v=ToTpfJ09_qs

Video: *God and Evolution: The Problem With Theistic Evolution* (1:02:35)
https://www.youtube.com/watch?v=9WHB_kMasMs&t=20s

Video: *The Problem with Theistic Evolution.* (9:04)
https://www.youtube.com/watch?v=GkAxRY41ndU

Video: Debate: *Did God Use Evolution?* With Dr. Karl Giberson (TE) and Dr. Randy Guliuzza (biblical creationist) (2:54:55) Sponsored by the Southern California Seminary Center for Creation Studies
https://www.youtube.com/watch?v=H9kl3S05X50

Trilobites

The Trilobite eye

Trilobite eyes are so complex that they are often referred to as a remarkable "evolutionary advance" in the development of sight, and their complexity "borders on... science fiction."

Trilobite Eyes, American Museum of Natural History (pro evolution)
https://www.amnh.org/our-research/paleontology/paleontology-faq/trilobite-website/the-trilobite-files/trilobite-eyes
This article showcases many images of diverse and complex trilobite eyes.

Video: *Trilobites and Punctuated Equilibria* with Niles Eldredge, 2015 (4:45) (pro evolution)
https://www.youtube.com/watch?v=9uqpcKWuhrU

The Trilobite Eye by S.M. Gon III (pro evolution)
http://www.trilobites.info/eyes.htm

Vestigial Organs

Vestigial organs are presented as either having no value or a reduced functionality from what they were originally.

Bergman, Jerry. *"Vestigial Organs" Are Fully Functional: A History and Evaluation of the Vestigial Organ Origins Concept.* (With George Howe, Ph. D., Forward by David Menton, Ph. D. Washington University School of Medicine; Preface by V. Wright M.D., F.R.C.P., Professor University of Leeds). Terre Haute, IN; Creation Research Society Books, 1990.

Audio: *There are no Vestigial Organs* with Dr. Jerry Bergman, 2017 (in 4 parts).
Part 1 (16:29) https://www.youtube.com/watch?v=64WSbc4eRWg
Part 2 (16:03) https://www.youtube.com/watch?v=HGyNmjYNzho
Part 3 (48:57) https://www.youtube.com/watch?v=HGyNmjYNzho
Part 4 (27:40) https://www.youtube.com/watch?v=7tP0Rwfwf4A

Appendix B

Video and Other Media Resource Providers

Answers In Genesis (AIG)
https://answersingenesis.org/
https://answersingenesis.org/media/video/

Access Research Network (ARN)
http://www.arn.org/
http://www.arn.org/news-videos/videos.html

Cornerstone Network-Origins
http://origins.ctvn.org/
http://origins.ctvn.org/past-episodes/

Creation in the 21st Century with David Rives
https://creationinthe21stcentury.com/
Video sample: https://www.tbn.org/programs/creation-21st-century-david-rives/watch/creation-21st-century-david-rives-38

Creation Ministries International (CMI)
https://creation.com/
https://creation.com/media-center

Discovery Institute
https://www.discovery.org/
https://www.discovery.org/id/

Institute for Creation Research (ICR)
https://www.icr.org/homepage/
https://www.icr.org/video
https://www.youtube.com/user/DiscoveryScienceNews

Illustra Media (since 1997) has produced some of the most compelling videos on a wide variety of subjects related to Intelligent Design issues. https://illustramedia.com/

Video: *Darwin's Dilemma (The Mystery of the Cambrian Fossil Record)* (1:13:57)
https://www.youtube.com/watch?v=xxh9o32m5c0

Video: *The Privileged Planet* (1:32:19)
https://www.youtube.com/watch?v=QmIc42oRjm8

Video: *The Case for a Creator with Lee Strobel* (59:47)
https://www.youtube.com/watch?v=ajqH4y8G0MI

Video: *Unlocking the Mystery of Life* (1:07:06)
https://www.youtube.com/watch?v=tzj8iXiVDT8
An early meeting of several key ID scientists at Pajaro Dunes, CA.

APPENDIX C

Apologists for Evidence of a Designed Universe

ARMITAGE, Mark[786]

Dr. Armitage has been working on some incredible research where he has been documenting the existence of non-mineralized soft-tissue cells he found in a fossilized Triceratops horn. To easily locate all of Mark's videos on You Tube, enter "MarkHArmitage" in the search bar.

If you would like to contribute to Mark's ongoing research, you can do so at www.gofundme.com/T5ZBHS

Video: *Report of Soft Tissues in a Triceratops Horn* (1:12:44)
https://www.youtube.com/watch?v=MqDV_MTQSxg

Video: *Mark Armitage Fired for Dino Soft Tissue Discovery...* (14:10) Excellent review by Dr. Armitage about his now famous and stunning discovery of soft tissue from a Triceratops horn and frill.
https://www.youtube.com/watch?v=vXCEMppadC4

Video: *Interview with Mark Armitage on Genesis Week* (42:31)
https://www.youtube.com/watch?v=h-8gjvNsOvM

Video: *Moments With Mark Interview Series*
Part 1 (23:03) https://www.youtube.com/watch?v=JHhChIEChw4
Part 2: (26:34) https://www.youtube.com/watch?v=dao5qrCM-_Q

Video: *Mark Armitage Walks the Plank for Dino Soft Tissue Discovery* (1:04:25)
https://www.youtube.com/watch?v=7OB2Obr9CDE

Audio broadcast interviewing Dr. Armitage, where he recounts how he found a Triceratops horn in Montana containing soft tissue. Armitage recounts details of his discovery, the publication of his results, the ridicule he received, and his termination from CSUN. He also talks about the similar sensational finding by paleontologist Mary Schweitzer. Since the recording of this interview, Armitage settled with CSUN out of court for religious discrimination.

---

786 Mark brings into focus one of the most sensational finds in paleontology: soft tissue in dinosaur bones. The presence of non-mineralized soft tissue and proteins in several dinosaur fossils may be the most significant paleontological discovery of the 21st century. This discovery directly challenges the evolution-biased timescale. Are dinosaur fossils really over 65 million years old? Can tissue and proteins survive in these fossils for over 65 million years?"

Video: *Young Earth Creationist wins lawsuit* – Dr. Armitage was fired from CSUN, and talks about that experience, his courtroom trial and the outcome of his lawsuit against them here:
Part 1: https://www.youtube.com/watch?v=JCq6KOLbdcE
Part 2: https://www.youtube.com/watch?v=We_XIq-k66c&pbjreload=10

For more videos about Dr. Armitage, including updates on what he's up to these days, go to this link:
https://www.youtube.com/user/MarkHArmitage

Armitage, Mark. 2016. "Preservation of Triceratops horridus Tissue Cells from the Hell Creek Formation, MT," *Microscopy Today*, January. pp: 16-22. https://www.cambridge.org/core/journals/microscopy-today/article/preservation-of-triceratops-horridus-tissue-cells-from-the-hell-creek-formation-mt/11CD094ED1312B6C618E098C12FCC324

_____, and Kevin Anderson. 2013. "Soft Sheets of fibrillar bone from a fossil of the supraorbital horn of the dinosaur Triceratops horridus." *Acta Histochemica* 115 603–608. https://www.sciencedirect.com/science/article/pii/S0065128113000020

AXE, Dr. Douglas (Director of Biologic Institute)

Video: *UNDENIABLE: How Biology Confirms Our Intuition That Life Is Designed* (1:29:56)
https://www.youtube.com/watch?v=SC9Hx3WpsCk

Video: *Biologist Douglas Axe on Challenges to Darwinian Evolution* (3:24)
https://www.youtube.com/watch?v=2grcHPo8oDQ

BECHLY, Günter [787]

Video: *A German Scientist Speaks Out about Intelligent Design* (6:02)
https://www.youtube.com/watch?v=fqiXgtDdEwM

Video: *The Fossil Record: Implications for Neo-Darwinian Evolution* (2:52:00)
https://www.youtube.com/watch?v=KcT61jEnJF8

---

787 To learn more about what happened to Günter Bechly, see Chapter 21 of *Censoring the Darwin Skeptics* (Volume III of the Slaughter of the Dissidents trilogy), Leafcutter Press, 2018.

Video: *Fossil Discontinuities: A Refutation of Darwinism and Confirmation of Intelligent Design* (53:18)
https://www.youtube.com/watch?v=M7w5QGqcnNs

BEHE, Dr. Michael J.

Audio: *Darwin's Black Box - A Debate - Michael Behe vs Keith Fox* (1:20:47)
https://www.youtube.com/watch?v=luRGzVrr2Cs

Video: *The Evidence of Design from Biology* with Dr. Michael Behe at the University of Toronto, November 16, 2012 (53:29)
https://www.youtube.com/watch?v=s6XAXjiyRfM&feature=youtu.be

Video: *Evidence of Design from Biology* (53:20)
https://www.youtube.com/watch?v=s6XAXjiyRfM

Video: *Revolutionary: Michael Behe and the Mystery of Molecular Machines* (59:55)
https://revolutionarybehe.com/

Video: *Unlocking the Mystery of Life* (1:07:06)
https://www.youtube.com/watch?v=tzj8iXiVDT8

Video: *The Creation Conversation - Part 1* - Stephen Meyer and Michael Behe (1:01:11)
https://www.youtube.com/watch?v=k-fVpctlERU&t=77s

Video: *From the Big Bang to Irreducible Complexity* (58:17)
https://www.youtube.com/watch?v=vL8KAJkdF2I

Video: *Should Intelligent Design Be Taught As Science? - Michael Behe vs Stephen Barr* (1:11:44)
https://www.youtube.com/watch?v=knEY1wKODR0

BERGMAN, Dr. Jerry

Video: *Slaughter of the Dissidents* (52:06)
Seattle Creation Conference (September 2013)
http://www.nwcreation.net/videos/Slaughter_of_the_Dissidents.html

Video: *Slaughter of the Dissidents* (26:34)
https://www.youtube.com/watch?v=MJ08osyl2EE

Video: *Slaughter of the Dissidents* (39:11)
Part 1 of 2 https://www.youtube.com/watch?v=NMBbtJt6_UU
Part 2 of 2 https://www.youtube.com/watch?v=sUExU_nKFCo

Video: *The Fossil Record* (26:31)
https://www.youtube.com/watch?v=hChM__WYevc

Video: *Darwin's Blunders, Frauds and Forgeries* (1:09:52)
https://www.youtube.com/watch?v=tFbJyF4e1bQ

Video: *The Fossil Record* (26:21)
https://www.youtube.com/watch?v=hChM__WYevc

Video: *Dan Barker vs Jerry Bergman - Does God Exist* – 2014 (1:39:17)
https://www.youtube.com/watch?v=9WT1B0RA5uQ

Video: *Hitler's Darwinian Worldview* (26:31)
https://www.youtube.com/watch?v=7COoeXO0GiQ

Video: *Mutations Debunk Darwin's Evolution.* (35:44)
https://www.youtube.com/watch?v=b_Acfw1RN0c&feature=youtu.be

BERLINSKI, Dr. David

Video: *The Link between Evolution, Science, and Progressivism* (13:53)
https://www.youtube.com/watch?v=2VJVGV8zAS4
Appearing on FOX News channel's Life, Liberty and Levin.

Video: *The Devil's Delusion* (1:03:29)
https://www.youtube.com/watch?v=vtgV2VP9iEQ

Video: *The Devil's Delusion: Atheism and Its Scientific Pretensions* (1:14:56)
https://www.youtube.com/watch?v=0XIDykeZplU

Video: *Dr. David Berlinski Rebellious Intellectual Defies Darwinism.* (1:21:54)
https://www.youtube.com/watch?v=KEXnq_tcM7c
Berlinski is not a Christian.

Video: *Atheism and its Scientific Pretensions* (42:11)
https://www.youtube.com/watch?v=FyxUwaq00Rc

Video: *David Berlinski Explains Problems With Evolution* (37:42)
https://www.youtube.com/watch?v=Z6ElA0--JNg&t=12s

Video: *Atheism Poisons Everything* (debate with Christopher Hitchens) (1:02:57)
https://www.youtube.com/watch?v=Z6ElA0--JNg&t=12s

Video: *Flaws in Darwin's Theory of Evolution* – exchange between Dr. David Berlinski and Eugenie Scott, circa 1997 (9:50)
https://www.youtube.com/watch?v=64VEM9IDxZ4

BIRD, Wendell R.

The Origin of Species Revisited: The Theories of Evolution and of Abrupt Appearance (2 vols.) 1989. New York, NY: Philosophical Library.

Be sure to read Philip Johnson's review of Bird's work in Constitutional Commentary (1990) Vol. 7, 427-443. This impressive work contains an excellent assemblage of evidence from some of the most reputable science practitioners and scholars across many disciplines, establishing the notion that abrupt appearance is a powerful and persuasive argument (except, of course, for the US Supreme Court) as an alternative to evolution. The fossil record shows fully-formed, abrupt appearance of all life forms, and without any evidence of the evolutionary precursors required by naturalistic evolution. Evolutionists typically attribute this lack of evidence to things such as rapid evolutionary development not captured in the fossil record, adding that the speed of evolution precluded the appearance of evolutionary linkages in fossil layers. But this does not account for the equally disturbing (for Darwinians) fact that after their sudden appearance, this is followed by no change (i.e., "stasis') in the structure of organisms found in the fossil record for the duration of their appearance. Both of these factual inconsistencies provide Darwinians with what should be extremely uncomfortable, unsettling and significant concerns about evolutionary predictions and presuppositions. Unfortunately, many evolutionists attempt to dismiss this conundrum with a variety of explanations that assume evolution in spite of what the fossil record shows, thus promoting evolutionary theory ahead and in spite of the evidence. These efforts are similar to trying to make square pegs fit into round holes even though it's very clear this is an ill-conceived task. This world class puzzle, even known to Darwin himself, has yet to be convincingly resolved today in the context of evolutionary development.

COLLINS, Francis

Video: *Why It's So Hard for Scientists to Believe in God* (4:36)
https://www.youtube.com/watch?v=pINptKQYviQ

DANA-BASHIAN, David

Video: *16 Nearly Impossible Issues for Evolutionists to Answer* is a public lecture given by David Dana-Bashian on July 9, 2013 in Garden Grove, California
https://www.youtube.com/watch?v=seGBTsLpMdI&t=1s

DEMBSKI, William

Video: *Conservation of Information and the End of Materialism*, March 2014. (44:08)
https://www.youtube.com/watch?v=7nIj5RpzIn8

Note: "On September 23, 2016, Dembski officially retired from intelligent design, resigning all his "formal associations with the ID community, including [his] Discovery Institute fellowship of 20 years." Source: Wikipedia

Video: *Detecting Design in Biology* - Dr. William Dembski and Dr. Niall Shanks (1:51:30)
https://www.youtube.com/watch?v=bQ4mW4wm4s8

Audio: *Intelligent Design vs. Evolution* – Dr, William Dembski and Michael Shermer (59:49)
https://www.youtube.com/watch?v=kVbXU4HsIT0&t=9s

Video: *Information and the End of Materialism* (44:08) – March 2014
https://www.youtube.com/watch?v=7nIj5RpzIn8

DENTON, Michael

Video: *Privileged Species.* (32:14)
https://www.youtube.com/watch?v=VoI2ms5UHWg

Video: *An Interview With Michael J. Denton.* (32:47)
https://www.youtube.com/watch?v=B-Nh3RjZQiI&list=PL3nZoKKpj8RL-u5lmCcE0nMjqcDLFrygz

D'SOUZA, Dinesh

Video: *How Do I Know God Exists?* (42:47)
https://www.youtube.com/watch?v=1Ien2Ah3lEY
(Viewing this video is highly recommended – KHW)

GENTRY, Dr. Robert V.

Video: *The Young Age of the Earth* (57:28)
Dr. Robert Gentry discusses the implications of his research on radiohalos in coal.
https://www.youtube.com/watch?v=5bTLuyCd9G8

Gentry, Robert V. 1992. *Creation's Tiny Mystery.* Knoxville, TN: Earth Science Associates, 3rd Edition.
Appendix: Gentry responds to Gary Brent Dalrymple's letter to Kevin Wirth
Here Gentry responds to veteran USGS geologist and radiometric dating expert Brent Dalrymple's criticisms of Gentry's claims about the nature of polonium halos and the problems they cause for conventional science. Dalrymple was an expert witness during the Arkansas trial of 1981. Gentry claims these halos are evidence of a young earth, Dalrymple opposes this view.
http://www.halos.com/book/ctm-app-19.htm

Dalrymple Arkansas Trial Commentary

Dalrymple's Arkansas trial deposition testimony:
http://www.antievolution.org/projects/mclean/new_site/depos/pf_dalrymple_dep.htm

Dalrymple's Arkansas trial testimony: http://www.antievolution.org/projects/mclean/new_site/pf_trans/mva_tt_p_dalrymple.html

Gentry Arkansas Trial Commentary

Gentry's Arkansas trial deposition testimony:
http://www.antievolution.org/projects/mclean/new_site/depos/pf_gentry_dep.htm
Gentry's trial testimony: not provided. It seems that full trial transcripts were provided for pro-evolution expert witnesses at the Arkansas trial of 1981, but no transcripts of the court testimony are provided from witnesses on the cre-

ation side.[788]

Gentry, Robert V., 1974, "Radioactive Halos in a Radiochronological and Cosmological Perspective," *Science*, Vol. 184, pp. 62-66.

Baillirul, Thomas A. 2001. *"Polonium Halos" Refuted. A Review of "Radioactive Halos in a Radio-Chronological and Cosmological Perspective,"* by Robert V. Gentry. Talk Origins Archive. http://www.talkorigins.org/faqs/po-halos/gentry.html#Gentry1992

Gentry, Robert V., Warner H. Christie, David H. Smith, J.F. Emery, S.A. Reynolds, and Raymond Walker, 1976, "Radiohalos in Coalified Wood: New Evidence Relating to the Time of Uranium Introduction and Coalification," *Science*, Vol. 194, pp. 315-318

GULIUZZA, Dr. Randy

Video: *Five Minutes with a Darwinist* (26:30) - using a memory tool called FLUFF
Focus the discussion
Less than persuaded
Unobserved important events
Failed mechanism for design
Freedom found in Creation Science
If you only had five minutes with a Darwinist, this is a proposed approach to a discussion designed to challenge Evolution.
http://origins.ctvn.org/2018/07/30/five-minutes-with-a-darwinist-2/

GISH, Dr. Duane

Video: *Evolution: Challenge of the Fossil Record*
https://www.youtube.com/watch?v=QDBPIq_Ihpo

Video: *Origin of the Universe*
https://www.youtube.com/watch?v=TPWI3Sx4IHI

---

788 McLean v. Arkansas Documentation Project. http://www.antievolution.org/projects/mclean/new_site/index.htm#Plaintiff's transcript

HOVIND, Kent

Video: *Lies in the Textbooks* (2:47:48)
https://www.youtube.com/watch?v=n_OlX7M5MLA

JEANSON, Dr. Nathaniel T.

Video: *GENETICS – The Riddle of 6,000 Years: Genetic Clocks Confirm Recent Creation* (45:06)
https://www.youtube.com/watch?v=96odokh9Ua0

JOHNSON, Dr. Philip

Book: *Darwin on Trial*. 1991. Regnery Gateway Publishing Co, Just for fun, look on page 115 under RESEARCH NOTES.
http://maxddl.org/Creation/Darwin%20On%20Trial.pdf

Audio: *What the Evolution Controversy is Really About* – 1999 (1:31:06)
https://www.youtube.com/watch?v=8i-AIFOS-v4

Video: *Unlocking the Mystery of Life* (1:07:06)
https://www.youtube.com/watch?v=tzj8iXiVDT8

Video: *Darwinism: Science or Philosophy?* With Dr. Philip Johnson (1:38:24)
https://www.youtube.com/watch?v=fJPUwtNkxEQ

Video: *Darwinism: Science or Naturalistic Philosophy?* Philip E. Johnson vs William Provine, Stanford University, April 30, 1994 (1:46:32)
https://www.youtube.com/watch?v=m7dG9U1vQ_U

Video: *Can Science Know the Mind of God? The Case Against Naturalism* (57:57)
https://www.youtube.com/watch?v=JUeOaPLQoK4&list=PL3nZoKKpj8RKT_cOAsrYAwtTl6Izg0wyY

Video: *Johnson on Darwinism* (57:57)
https://www.youtube.com/watch?v=ww6T8xjp9Vo&list=PL3nZoKKpj8RKT_cOAsrYAwtTl6Izg0wyY&index=2
(Produced by Art Battson of Access Research Network)

Video: *Intellectual Breakthrough* (54:16)
https://www.youtube.com/watch?v=8qCNwBit4yk&index=3&list=PL3nZoKKpj8RKT_cOAsrYAwtTl6Izg0wyY

Video: *Darwinism on Trial* (1:29:47)
https://www.youtube.com/watch?v=-8meWGZ_e_Y

Video: *Blind Watchmaker? A Skeptical Look at Darwinism* (1:12:55)
https://www.youtube.com/watch?v=p2MwUgi8dlc&t=338s

KENYON, Dr. Dean H.

(Kenyon is a former evolutionist who changed his mind and now supports intelligent design Wrote the book Biochemical Predestination, an evolutionary classic)

Video: *Unlocking the Mystery of Life* (1:07:06)
https://www.youtube.com/watch?v=tzj8iXiVDT8

Video: *On the Origin of Life* - An Interview with Dr. Dean Kenyon (53:39)
https://www.youtube.com/watch?v=uJDa9QLP4aE

Video: *Focus on Darwinism* – An Interview with Dean Kenyon (36:25)
https://www.youtube.com/watch?v=KLpoAdUptas&t=6s

Video: *The Origins of Life* (57:59)
https://www.youtube.com/watch?v=ISIf58X4aBQ

LENNOX, John

Video: John Lennox: *The Question of Science and God*
Part 1 (47:33)
https://www.youtube.com/watch?v=gDjNv-ea56E

Part II (58:27)
https://www.youtube.com/watch?v=Tr3ghb6JG6A

Video: Richard Dawkins vs John Lennox | *Has Science Buried God?* Debate (1:21:52)
https://www.youtube.com/watch?v=OVEuQg_Mglw

Video: Richard Dawkins vs John Lennox | *The God Delusion Debate* (1:46:39)
https://www.youtube.com/watch?v=zF5bPI92-5o

Video: *John Lennox Reflects on Stephen Hawking's Life & Beliefs* (24:44)
https://www.youtube.com/watch?v=If4XisIJNA4

Video: John Lennox's question and answer session on '*Are God and Faith Anti-*

Science and Anti-Reason?' (38:30)
https://www.youtube.com/watch?v=GSu3p3wvtrs

Video: *Is Faith Delusional?* John Lennox, Ravi Zacharias (47:04)
https://www.youtube.com/watch?v=6ktJtmaq_Ok

LISLE, Dr. Jason

Video: *The Ultimate Proof of Creation* (1:33:33)
https://www.youtube.com/watch?v=uwXyvQinQyo

Video: *The Physical World Obeys God's Math (Fractals)* (1:01:52)
https://www.youtube.com/watch?v=NaDvPeWjBuY

Video: *Distant Starlight In a Young Universe* with David Rives and Jason Lisle (27:32)
https://www.tbn.org/programs/creation-21st-century-david-rives/watch/creation-21st-century-david-rives

LUSKIN, Casey

Luskin, Casey. 2009. "Does Challenging Darwin Create Constitutional Jeopardy? A Comprehensive Survey of Case Law regarding the Teaching of Biological Origins." *Hamline Law Review*. Vol. 32 (1)1, Winter.
https://www.discovery.org/a/11291/
This article attempts to exhaustively survey the case law relevant to the teaching of biological origins, dividing the cases into three major categories: (1) Cases upholding the right to teach about evolution; (2) Cases rejecting the teaching of alternatives to evolution; and (3) Cases rejecting disclaimers regarding the teaching of evolution. The range of constitutionally permissible policies for teaching evolution can also be understood by studying policies that have not engendered lawsuits. Twenty-one cases are reviewed, as well as various policies that have not faced legal challenges, revealing that while courts have firmly upheld the rights of educators to teach evolution and have rejected attempts to teach creationism, none of these cases stands for the proposition that a curriculum that teaches scientific critiques of evolution would necessarily place a school board in constitutional jeopardy.

Luskin, Casey. 2015. M.S., J.D., ESQ. Darwin's Poisoned Tree: Atheistic Advocacy and the Constitutionality of Teaching Evolution in Public Schools." *Trinity Law Review* Fall, Volume 21, Issue 1, Pp. 130-233. Published by Trin-

ity Law School
https://www.discovery.org/m/2019/03/Darwins-Poisoned-Tree-Casey-Luskin-Trinity-Law-Review-Spring-2016.pdf

MALONE, Bruce

Videos:

Episode 1: *Evolution vs. Creationism: Were Isaac Newton & Albert Einstein right?* (52:34)
https://www.youtube.com/watch?v=gUaNyW8owXw&list=PL6NNBo_y_fjPfDQRSyCYO8DSEqDyGfvUe&index=1

Episode 2: *Natural Selection, DNA Research, Spontaneous Life Generation* (53:45)
https://www.youtube.com/watch?v=v7_kXHNZX-o&index=2&list=PL6NNBo_y_fjPfDQRSyCYO8DSEqDyGfvUe

Episode 3: *Dinosaurs, Fossils, Young Earth, & The Ice Age* (55:03)
https://www.youtube.com/watch?v=HNnfT4D0Z9c&list=PL6NNBo_y_fjPfDQRSyCYO8DSEqDyGfvUe&index=3

Episode 4: *The War Between Science and Religion* (54:50)
https://www.youtube.com/watch?v=jF0ztZtoK80&list=PL6NNBo_y_fjPfDQRSyCYO8DSEqDyGfvUe&index=4

Video: **Explosive Geological Evidence for Creation** (1:13:21)
https://www.youtube.com/watch?v=kNoXHP5OE14

Mc LATCHIE, Jonathan

Apologetics Academy
http://apologetics-academy.org/
https://www.youtube.com/channel/UCr6nvYFyoCk5DsRgYRrhC_g/videos

Video: *The Waiting Time Problem in Evolutionary Biology* (3:45)
https://www.youtube.com/watch?v=iLfmfmI2HvM

MENTON, Dr. David

Video: *Evolution, not a Chance!* (53:40)
https://www.youtube.com/watch?v=Jck4naOUGQo

Video: *Fossil Missing Link Fraud: LUCY* (1:11:48)
https://www.youtube.com/watch?v=QDBPIq_Ihpo

MEYER, Dr. Stephen C.

Meyer, Stephen C., "DNA and the Origin of Life: Information, Specification, and Explanation" [date unknown] 44 pp.
www.vediciIluminations.com/downloads/Intelligent-Design/DNAPerspectives.pdf
From the abstract: "the term 'information' can designate several theoretically distinct concepts. By distinguishing between specified and unspecified information, this essay seeks to eliminate definitional ambiguity associated with the term 'information' as used in biology."

Video: *Intelligent Design 3.0 – Stephen C. Meyer* (2018) (1:36:48)
https://www.youtube.com/watch?v=lgs6J4LqeqI&feature=youtu.be

Video: *God & Evolution: A Critique of Theistic Evolution* (2018) (2:23:43)
https://www.youtube.com/watch?v=bMMZ48M1TOo

Video: *Return of the God Hypothesis* – (2014) (50:49)
https://www.youtube.com/watch?v=iKBv7qgyztc&feature=youtu.be

Video: *Darwin's Doubt* (1:25:47)
https://www.youtube.com/watch?v=QiDmtDuMHSc

Video: *What is Intelligent Design?* (1:00:45)
https://www.youtube.com/watch?v=jJIbcE0kOAs

Video: *The Failure of Darwin's Theory* (28:30)
https://www.youtube.com/watch?v=pZyRgYZe6tM&feature=youtu.be

Podcast: *Cross-Examining Dr. Stephen Meyer* (2009) (35:48)
Hosted by John West of Discovery Institute
https://www.youtube.com/watch?v=qEJVxUVEMxk

Video: *What's Behind it all? God, Science and the Universe.* (2:26:46)
Krauss, Meyer, Lamoureux
Wycliff College at the University of Toronto – March 19, 2016
https://www.youtube.com/watch?v=mMuy58DaqOk&feature=youtu.be

Video: *Unlocking the Mystery of Life* (1:07:06)
https://www.youtube.com/watch?v=tzj8iXiVDT8

Video: *Peter Ward vs Stephen Meyer - Evolution vs Intelligent Design* (1:33:31)
https://www.youtube.com/watch?v=01P4py7NUMc

Video: *The Evidence of the Fossil Record*. (with Paul Nelson and Jonathan Wells) (28:30)
https://www.youtube.com/watch?v=R4yGdJ0DOgM

Video: *Signature in the Cell*. Meyer Faces His Critics.
Part 1 (1:21:02)
https://www.youtube.com/watch?v=eW6egHV6jAw
Part 2 (1:01:23)
https://www.youtube.com/watch?v=OQf29Pden30

Video: *Rock of Ages & the Ages of Rocks* - Stephen Meyer
The Christian Mind - Part 7 (1:03:04)
https://www.youtube.com/watch?v=dvMQXzidVG4&t=8s

Questions and Answers – Part 8 (1:16:26)
https://www.youtube.com/watch?v=ksFL0roNnIE

Video: Darwin: *A Myth for the Post-Christian Mind* (47:17)
https://www.youtube.com/watch?v=PbcY9iya40o&t=5s

Video: *The Creation Conversation - Part 1* - Stephen Meyer and Michael Behe (1:01:11)
https://www.youtube.com/watch?v=aM0KcIt_pXw

Video: *M. Shermer, D. Prothero vs S. Meyer, R. Sternberg - The Origin Of Life –* 2009 (1:55:00)
https://www.youtube.com/watch?v=7yqqlZ29gcU

Video: *Amazing Seminar on Intelligent Design by Stephen Meyer* (1:22:00)
https://www.youtube.com/watch?v=b7Vf6MvBiz8&feature=youtu.be

Curriculum Vitae: http://www.stephencmeyer.org/curriculum-vitae.php

For a critical review of Meyer's work, Signature in the Cell, please refer to Intelligent Design, Abiogenesis, and Learning from History: A Reply to Meyer by Dennis R. Venema, Perspectives on Science and Christian Faith. Volume 63, Number 3, September 2011.
https://www.asa3.org/ASA/PSCF/2011/PSCF9-11Venema.pdf

MILLER, Kenneth R.

Video: *The Collapse of Intelligent Design* (1:58:41)
https://www.youtube.com/watch?v=Ohd5uqzlwsU&t=421s

Video: *Debate - Kenneth Miller vs Paul Nelson -* Evolution vs Creationism – 2005 (1:17:31)
https://www.youtube.com/watch?v=px5-uG2wFZI

MILTON, Richard

Video: *Forbidden Science – Shattering the Myths of Darwin's Theory of Evolution* (32:33)
Excellent video of this journalist's research and experience related to this issue. He has also published a book under this title. This very candid and riveting interview was filmed for the documentary The Mysterious Origins of Man: Rewriting Human History. He articulates problems with dating methods and lack of evidence for evolution.
https://www.youtube.com/watch?v=7Wr-lXLGCxQ&feature=youtu.be

MINNICH, Dr. Scott

Video: *An interview with Scott Minnich.* (29:36)
https://www.youtube.com/watch?v=tm-Ukz72AdA

MOHLER, Dr. Albert

Video: *Why Does the Universe Look So Old?* (1:05:45)
https://www.youtube.com/watch?v=ggJZz3WkTCI&t=43s

Video: *Dinosaurs* (9:15)
https://www.youtube.com/watch?v=I_Wi5OYZ7Ks

Video: *Creation Debate: Dr. R. Albert Mohler, Jr. and Dr. Jack Collins* (2:13:26)
https://www.youtube.com/watch?v=kGETfOQgNI4

MORRIS, Dr. Henry

Video: *The Genesis Flood* (54:10)
https://www.youtube.com/watch?v=CTG472F4dJY

Video: *The Dark History of Evolution* (47:13)
https://www.youtube.com/watch?v=5hk09uoDgm0

Video: *The Troubled Waters of Evolution* (5 parts)
https://www.youtube.com/watch?v=1juOQvOdvQ8&list=PL96318E6C60A5D7CE

Audio: *The Long War Against God* (55:02)
https://www.youtube.com/watch?v=BkpoFv-D6L4

Audio: Debate: *The Theory if Evolution is Superior to the Theory of Special Creation as an Explanation for all the Scientific Evidence Related to Origins* with Dr. Henry Morris and Ken Miller. (3:10:21) April 10, 1981.
https://www.youtube.com/watch?v=EDZ_qdEB39Q

MORRIS, Dr. John

Video: *Dragons or Dinosaur* with Dr. John Morris, Darek Isaacs, Dr. David Menton, and Chuck Missler (1:24:27)
https://www.youtube.com/watch?v=gJYqu35jtxA

Video: *Fossil Record: A Problem for Evolution* (1:04:24)
https://www.youtube.com/watch?v=aESGdORhHlg

NELSON, Dr. Paul

Video: *Unlocking the Mystery of Life* (1:07:06)
https://www.youtube.com/watch?v=tzj8iXiVDT8

Video: *The Genius of Birds* (26:32)
http://origins.ctvn.org/2018/02/19/the-genius-of-birds-2/
Dr. Nelson has a Ph.D. in Philosophy of Biology from the University of Chicago. In this video Dr, Nelson discusses the genius of bird design. One of the key questions he addresses is: "How can you deny the scholarly consensus that birds evolved from theropod dinosaurs by an undirected evolutionary process?"

Video: *The Butterfly Enigma* (26:32) Why butterflies create an incredible challenge for evolutionists.
http://origins.ctvn.org/2018/08/06/the-butterfly-enigma-2-2/

Video: *What Happened to the Tree of Life?* (1:08:13)
https://www.youtube.com/watch?v=Mf01GXUl-mg

Video: *Molecular Machines and the Death of Darwinism* [Dembski, Wells, Nelson, Macosko] (43:14)
https://www.youtube.com/watch?v=-D2ng2mvyes&t=7s

Video: *Debate - Kenneth Miller vs Paul Nelson -* Evolution vs Creationism – 2005 (1:17:31)
https://www.youtube.com/watch?v=px5-uG2wFZI

Video: *Evolution is a Theory of Transformation, not Similarity* (2:30:21)
https://www.youtube.com/watch?v=K1b5kW-ioeo

OLIVER, Rick

Video: *Microbiologist Loses Faith in Evolution* (59:58)
https://www.youtube.com/watch?v=46OzGYqpfkk

PATTON, Dr. Don

Video: *Lack of evidence for Evolution: The Fossil Record Supports Creation.* (1:09:25)
https://www.youtube.com/watch?v=gSof6z_Noqs&t=1s

Video: *The Creation Model vs. Evolution Model ("What is Creation Science?")*
https://www.youtube.com/watch?v=cxMkMBXAVZ8

REMINE, Dr. Walter

Video: *The Biotic Message* (2:19:22) Lecture on "Message Theory," an alternative to macroevolution.
https://www.youtube.com/watch?v=o2USDd_8eAY

Video: *An Interview With Dr. Walter Remine* (1:26:54)
https://www.youtube.com/watch?v=mj2f9Py4vEQ

ROSS, Dr. Marcus

Video: *Rocks as Clocks* (26:30)
http://origins.ctvn.org/2017/09/03/rocks-as-clocks-3/

Video: *Noah's Flood: Where Genesis Meets Geology*, Evangel University Faith & Science Conference, September 23, 2016 Discussion on Plate Tectonics and the totality of the Flood.
https://www.youtube.com/watch?v=5wlwm1COQZw

Video: *What Does the Fossil Record Say?* with Kevin Conover 2017 (57:20)
https://www.youtube.com/watch?v=4KvXcYpCDdU

SANFORD, John

Video: *How Evolution Hurts Science* (1:01:18)
https://www.youtube.com/watch?v=P-H4X2b7x7Q&feature=youtu.be
Sanford argues that the true nature of the debate is actually defined in the two dimensions of Faith vs. Faith and Science vs. Science. Sanford exposes 5 faith-based positions of secular evolutionists.

Video: *John Sanford on Genomic Entropy* (1:40:32)
https://www.youtube.com/watch?v=_edD5HOx6Q0

SCHAEFER, Dr. Henry

Video: *The Kalam Cosmological Argument* (53:45)
https://www.youtube.com/watch?v=Ziojymi5OYY

SCHWEITZER, Mary

Video: *60 Minutes expose on Mary's T-rex soft dinosaur tissue discovery*
https://www.youtube.com/watch?v=yJOQiyLFMNY

Video: News segment interview with Mary Schweitzer
https://www.youtube.com/watch?v=ynXwAo9V_pY
Note: Mary maintains her findings exist in an evolutionary context, however her research has obvious implications for those who advocate a recent creation.

SNELLING, Andrew

Video: *Radioactive and Radiocarbon Dating*
Part 1: https://www.youtube.com/watch?v=m6l-vnBvZj0 (15:25)
Part 2: https://www.youtube.com/watch?v=qc_1-4FX368 (15:18)
Part 3: https://www.youtube.com/watch?v=JshYhKpLMS8 (15:18)

Part 4: https://www.youtube.com/watch?v=LPVJbRaQkNI (15:15)

Video: *Radioactive and Radiocarbon Dating: Turning Foe Into Friend* - Dr. Andrew Snelling (1:01:44)
https://www.youtube.com/watch?v=JIZo4o77kRI

STERNBERG, Dr. Richard

Video: *Whale Evolution vs. Population Genetics* with Dr. Paul Nelson (11:07)
https://www.youtube.com/watch?v=0csd3M4bc0Q

M. Shermer, D. Prothero789 vs S. Meyer, R. Sternberg - *The Origin Of Life* – 2009 (1:55:00)
https://www.youtube.com/watch?v=7yqqlZ29gcU

THAXTON, Dr. Charles

Video: *Focus on the Origin of Life* (1:13:03)
https://www.youtube.com/watch?v=5AXkrc2OSs4&t=196s

THOMAS, Brian.

Video: *Dinosaurs, Soft Tissue, and the Bible - Unlocking the Mysteries of Genesis* (1:08:29)
https://www.youtube.com/watch?v=FOG39eITVyo

TOUR, James.

A practicing scientist (synthetic organic chemist) and professor doing work on the synthesis of single-molecule nanomachines, which includes molecular motors and nanocars. Tour developed the *NanoKids* concept for K-12 education in nanoscale science, and also *Dance Dance Revolution* and *Guitar Hero* science packages for elementary and middle school education: *SciRave (*www.scirave.org*)* which later expanded to a Stemscopes-based *SciRave*. The *SciRave* program has risen to be the #1 most widely adopted program in Texas to complement science instruction, and it is currently used by over 450 school districts and 40,000 teachers with over 1 million student downloads.

---

789 Video: *Evolution: What the Fossils Say* by Donald Prothero.
https://www.youtube.com/watch?v=DjFgcOId-ZY

https://www.jmtour.com/
To read his extensive vitae, go to https://www.jmtour.com/publications/

Video: 2019. *The Mystery of the Origin of Life* (58:01)
https://www.youtube.com/watch?v=zU7Lww-sBPg&feature=youtu.be
Taped at the 2019 Dallas Science and Faith Conference at Park Cities Baptist Church in Dallas. Sponsored by Discovery Institute's Center for Science and Culture. In this video, Dr. Tour discusses the impressive use of his work in various industries and applications. Starting at 8:00 into the video he launches into a discussion of "what does science tell us about the origin of life?" without "making any reference to God or intelligent design." He describes the complexity of cells and makes the point that molecules don't care about, assemble or evolve non-regular patterns of organic material into living systems. This video presents an excellent review of the difficulties surrounding the non-directed (i.e., evolutionary) formation of cellular components.

VEITH, Walter.

(Former atheist and evolutionist who converted to Christianity)
Video: *The World According to Darwin*
https://www.youtube.com/watch?v=leeS3I_jNGo

Video: *The World's Strongest Evidence for Creationism* (1:33:05)
With Dr Ben Carson
https://www.youtube.com/watch?v=gFDIriIhbIM

WELLS, Jonathan

Video: *Icons of Evolution by Jonathan Wells* (1:50:39)
https://www.youtube.com/watch?v=te3aShKST1A

Videos: *Evolution vs. Intelligent Design* Jonathan Wells and Michael Shermer (7 parts)
https://www.youtube.com/watch?v=2Pm9FC1S3Xs

Video: *Evolution Exposed with Jonathan Wells* (13:35)
https://www.youtube.com/watch?v=YtB_-VpIJrc

Video: *God and Evolution - Scientific Critique of Francis Collins› BioLogos* (25:52)
https://www.youtube.com/watch?v=0NIJ0NdW_Ho

Video: *Science and Theistic Evolution* (25:53)
https://www.youtube.com/watch?v=hksGZcqJ5h4

WILE, Dr. Jay L.

Video: *Debate: Creationism vs. Evolution* (1:24:51)
https://www.youtube.com/watch?v=nIeEcG6WRxo

About the Author
Jerry Bergman, Ph. D.

Dr. Bergman has 9 earned degrees, including 5 graduate degrees and a Ph.D. from Wayne State University in Detroit, Michigan. His over 1,000 publications are in both peer- reviewed, scholarly and popular science journals. Dr. Bergman's work has been translated into 13 languages including French, German, Italian, Spanish, Danish, Polish, Arabic, and Swedish.

The 1,026 college credit hours he has earned is the equivalent to almost 20 master's degrees. According to the website *The 10 Most Educated People on the Planet*, Bergman would fit in the list ahead of several other people. Assuming the accuracy of the list, he thus appears to be one of the most formally educated persons in the world. If those in the science area were separated, Bergman would be first on the list.

His books and/or books that include chapters he has authored are in over 1,400 college and major public libraries in 26 countries. So far over 80,000 copies of the 43 books and monographs that he has authored or co-authored are in print. He has spoken over 2,000 times to college, university, and church groups in America, Canada, Europe, Africa and Asia. He has also been a guest on hundreds of radio and television shows.

About the Publisher
Kevin H. Wirth

Kevin H. Wirth launched Leafcutter Press in 2008, publishing modestly since that time. Perhaps the most significant project he oversaw was the *Slaughter of the Dissidents* trilogy by Dr. Jerry Bergman (2008-2018) where he served as both senior editor and contributor.

Prior to that time, he had previously worked as the Director of Research for Students for Origins Research (later Access Research Network) for a number of years, providing support for their quarterly publication *Origins Research*. While with SOR, he generated an extensive bibliography database of article and book resources on the topics of creation and evolution. He also used this to provide extensive background research for the launch of the groundbreaking book *Darwin on Trial* (1991) by Philip E. Johnson, Professor of Law Emeritus at the UC Berkeley School of law. In addition, he provided significant research support for the 2 volume work by Yale Law School graduate Wendell R. Bird titled *Origin of Species Revisited: the Theories of Evolution and of Abrupt Appearance* (1991). Listed in America's Premier Lawyers, Bird is a Visiting Scholar at Emory University School of Law and was the attorney who argued before the US Supreme Court (argued December 10, 1986 – decided June 19,1987) on behalf of the appellants in the Edwards v. Aguillard case (482 U.S. 578).

Wirth grew up in San Carlos, California, located in the San Francisco bay area, where he took advantage of the extensive Stanford University libraries for most of his research work. An avid photographer and naturalist, he currently resides in Spokane, Washington.

CPSIA information can be obtained
at www.ICGtesting.com
Printed in the USA
FSHW011348210220
67225FS